Dear To My Heart

Eugenia Irene

To
Sharon
Great lady

Eugenia Jones

phone: 334-1092

Canadian Cataloguing in Publication Data

Irene, Eugenia, 1930-
Dear to my heart

ISBN 0-919537-54-5

1. Yawney, William, 1905-1992. 2. Frontier and pioneer life--Alberta. 3. Farmers--Alberta--Biography. 4. Merchants--British Columbia--Biography. 5. Vernon (B.C.)--Biography. I. Title.

FC3849.V47Z49 2000971.23'02092 C00-910916-1
F1078.I73 2000

Photo Credits
front cover - W. Gibson Kennedy, courtesy of Kettle Valley Connection
back cover - Ray Matthews

Printed in Canada

DEDICATION

To the Memory of my Father
William Yawney

Whose kindness, strength, understanding
and love will be remembered forever

Why God Made Fathers

God knew that children all would need
Someone secure and strong
To shelter and protect them,
And to teach them right from wrong…

Someone to take pride in,
And look up to as a guide,
Someone they could count on,
And in whom they could confide…

He knew as children grew up,
They'd need the reassurance of
Someone with faith and trust in them,
Who would always give them love…
And that's why God made fathers.

Katherine Nelson Davis

ACKNOWLEDGEMENTS

With appreciation and thanks;

To my mother Katharine for the information here written of my father's real estate purchases 1928 – 1992.

To Anne for remembering all, or most of all, the nannies' names who worked in our home summer and winter 1929 – 1939.

To Edward for supplying family and farm pictures.

To Henry who at the age of nine years helped me chase five cows along our country road thirteen miles to the Ranfurly stock yard.

To Patrick, Marshall and Joyce for being family.

To Susan for her excellent typing of my hand written manuscript.
To Betty for her cooperation and understanding.

And to my dear precious son William Alexander Frederick. Your encouragement through the years gave me strength to go on.

I love you

To Barry Broadfoot, a thank you from the bottom of my heart.

My father always said " Whatever you choose to do, do it right or don't do it at all." Those words stayed within me through the years and when I began writing my book " Dear to my heart" I searched for and chose the very best for an opinion and a reading. Writing this book was extremely important to me and I was fortunate to have found the very best - Canada's most famous author - Barry Broadfoot, author of "The Pioneer Years: 1895-1914", "The Dirty Thirties, Ten Lost Years: 1929-1939", "Six War Years: 1939- 1945", "Years of Sorrow, Years of Shame.",and "Barry Broadfoot 1926." "My Own Years"- first edition 1983.

My first encounter with Barry Broadfoot, doctor of Law and Literature, was the summer of 1993 when I began writing my book. Conversation with him was exciting and most interesting. Conversation with him gave me the inspiration and motivation to write and to bring back some of the history of the pioneer years. Together with the life of a true Canadian; an only son of a pioneer; my father, 1905 - 1992.

September 1998, Barry Broadfoot said he read with criticism and interest my story title " Dear to my Heart" and it was then we met for the first time to discuss my works.

Eugenia Irene
Courtenay 2000

FOREWORD

During the 1800's and early 1900's all over Europe, talk about Canada was like a restless wind that refused to die down. According to pamphlets and leaflets, Canada sounded all right: free land, schools, meat, fish, fruit which grew practically by themselves the land was so rich. The Canadian Government wanted this pre-empted land settled and improved and allotted 160 acres – a quarter section of land to homesteaders.

Families; eager settlers from British and European countries unaware of the harsh demands of pioneer life were arriving in Montreal in freight and cattle ships under very poor conditions with their young families and infants. Canada was to be their land of hopes and dreams. Some of the immigrants arrived with money to buy a wagon, an ox, horses, a cow and food. Some came with nothing but their strength and eagerness to build a new life.

During the early 1900's, the pioneers travelling west from Montreal travelled by train to Saskatoon, the end of the railroad line where they camped a while, to rest a few days before travelling on. They faced a formidable overland trek by horse – or oxen-drawn wagons to Lloydminster a tent-city rest station – now a city on the Alberta Saskatchewan border. The townsite was named after Reverend Lloyd, who insisted on the appendage of "Minister". From Lloydminster, the pioneers who were unable to buy wagons, an ox or horses, walked to their homesteads. Pioneers who settled on the prairies built sod houses and the pioneers who were allotted forest acres built cottages of logs.

Some pioneers lived in cardboard box shelters. During the 1960's I worked in an office with an Irish lady who was born and lived in a cardboard box shelter until she was 8 years old.

Those who lived in that era are gone, with little left behind but unlabelled, tinted photographs of sober-faced people beside nondescript buildings, or rusted farm machinery in what is generously called a museum. Very few talked about, or wanted to remember, their life as pioneers. To them those years were best to be forgotten.

The pioneers did not want to remember those years; but we must, we must remember our brave, proud and strong pioneers who sacrificed their lives, homes and families to build a Canadian foundation for us, for our future generations.

9 year old William with his mother

Chapter 1

Michael and Annie somehow, reached their homestead in the far north in Alberta. Alberta was still the Northwest Territories in those days and that area was known as Buchach "The Solid Ukraine". This homestead was a forest of poplar trees, some birch and some evergreen spruce with a small lake near the entrance. With a handsaw, together they cut down trees to build their cottage of logs while they lived in a tent. With bare hands and strength they worked day and night to have their cottage built. Wet clay-soil was dug up from the lake's edge and mixed with dry grass. With this mixture they filled in the open spaces between the logs in the walls and the ceiling. The inside walls were then whitewashed.

Winter was soon upon them wrapping this land in a thick blanket of ice and snow. The long winter months were severe. The cold, frozen branches on the trees made snapping, crackling noises all day long. And through the cold winter nights, the only sounds heard were timber wolves howling and coyotes yelping all around them. There was no one there, no one anywhere near, they were all alone.

A curious Native Indian who was checking his trap lines once a month would always stop by to look them over and Michael and Annie would always invite him into their cabin to get warm and to have a cup of hot cocoa or tea. Somehow they made conversation, each in his own language and somehow they understood and accepted each other's company.

At the first sign of spring, Michael would leave this homestead to work on building the Canadian National

Railway approaching from Eastern Canada. Steel rails were being put down to run through Innisfree, Ranfurly, Vegreville then to Mundare and on to Edmonton. Leaving their homestead and Annie all alone with their three young daughters Maria, Julianna and Paulina, Michael would not be returning until the late autumn. With a shovel and a pick he was off to build a railroad that was to cross his Canada from sea to sea so that people soon arriving to join him and his family in Buchach would have a train to travel in all the way - and not in wagons over dusty trails.

1905

Early one morning as the sun's rays were slowly creeping over the still, calm, sleeping land: A lonely sad howl of a timber wolf echoed through the trees while in that cabin by the light of a coal oil lamp a new born baby cried. April 30, 1905, on this homestead north of Innisfree, in Alberta amongst rolling hills of poplar trees, wild roses, clear cold lakes of Whitefish and beavers dams, my father was born.

Without a whimper, not to waken her three little daughters asleep in their little straw beds, alone she delivered her precious son. Her name was Annie, she was of medium height with long dark hair, fair skin and a look of sadness in her curious hazel eyes. Her husband Michael was away building the Canadian Railway near Vegreville when the first springtime payday arrived and all the railroad men got paid in cash. Most of the men found a way to get to Edmonton to drink beer in the fancy hotel beer parlours. Some of the men went to Edmonton to see the city lights and to experience life in the city. And Michael; well! Michael was a gambler and there in Edmonton he gambled late into the night. He knew when to hold them, he knew when to fold them, and he knew when to walk away. When daylight came Michael was on his way to his cabin home. Over fields, hills and vales he went to be with Annie, to be at her side. With pockets full of money and the thrill and the happiness knowing a baby was due, Michael whistled as he walked and he sang a lullaby.

Колисала Мати Сина
Колисала Таїі Співала:

Ніц Надходи, Треба Спати
Коло Теби Рідна Мати
Ніц Надходи, Треба Спати
Коло Теби Рідна Мати.

Two men who watched Michael win big money followed him and in the wilderness many miles away from his home, the two men caught up to Michael. They beat him severely, they took all his money and they left him in a cold, wet gully to die. When he was found a few days later by a Native Indian trapper, he was very cold, tired, and he was ill. He had not slept a wink nor had he any food while he struggled to stay alive. He had to get out of that wilderness, to get to his home, to be with Annie. The Natives took Michael first to Vegreville to be checked by a doctor before they delivered him to his homestead cabin to be with his family. As Michael opened the cabin door, he stood there a while and he watched Annie holding their baby son and she was singing.

Я Тебе Нагодувала
І повила, І приспала.

He then came in, he embraced them both. With tears of happiness running down their faces, Michael kissed Annie and his son on the forehead. He then gently, tenderly, took the baby from the mother's arms, and he held him so close to him, he could feel his little heart beat. He looked into his son's dark eyes, and he saw a little smile on his baby's face. Michael then looked at Annie, and he said, " We will call him William". Annie said "yes" and "Michael". They named their son William Michael. With much tenderness and emotion Michael held his baby son and continued singing the lullaby Annie was singing when he opened the cabin door.

Колишу Тебе І Співаю
Спи Мій Сину
Баю Баю.
Колишу Тебе І Співаю
Спи Мій Сину
Баю Баю.

Days, then weeks were passing by and Michael's strength seemed to be rapidly fading. He was burning up with fever and was not able to return to the railroad camp. Annie alone plowed the field with an ox and seeded the grain; which was wheat and oats. She had cows to milk, chickens and geese to protect from coyotes and weasels, a vegetable garden to plant, and to nurture a new born baby son and her three small daughters. Maria the oldest, was ten years old when the baby was born. Annie's husband Michael required very much care, she tried nursing him back to health, though all the care she was able to give Michael together with help from the doctor in Vegreville, they could not save him. Michael's last words to Annie were "give our precious son the best care possible, it is through him that our name will live on forever on this our new Canadian home land". Michael's baby son William was six weeks old when at the age of thirty years old Michael died of pneumonia. He was buried in a handmade wooden casket on a little hill under his favourite tree, an Alberta spruce, with a wooden cross marking his grave. This was his Canadian homestead, amongst rolling hills of poplar trees, blossoms of beautiful wild Alberta pink roses growing everywhere from early spring on through the summer. And in a clear, cold lake to the left of the entrance on his homestead, were many whitefish and beaver dams.

Michael and Annie came to Northern Alberta, then called the Northwest Territories, knowing no one, or what life in Canada would be. They were all alone, like virtually every immigrant facing an unknown future but full of hope for a new good life. Michael's ancestors were known to be Ukrainians from the Jablonovy Mountains of Upper Mongolia, called an "Apple Valley." He was a big man with dark, dark eyes, black hair, a black mustache, and his skin was darkly tanned.

Annie was from the Krasnoyarsk area in the Yenisey River Valley, of Jewish and Ukrainian ancestry. Alone now in the wilderness with her baby son William and daughters, Maria, Julianna and Paulina; early one morning as she stood at her cabin window looking out over the fields where Michael was put to rest; she was praying and when she opened her eyes she said she saw Indians fishing in the lake on her land. She then took a big stick and ran out, yelling at them, to get off her land; and there she found they were catching fish for her, they knew she was alone with small children, and they were there to help her. The winter months that followed, the Native Indians brought her a deer at times and left it on her doorstep. Sometimes they brought some ducks and sometimes rabbits.

She was grateful for their help and the compassion they showed towards her and her children. In exchange for their kindness she gave them farm fresh milk for their babies.

The seasons came and the seasons went. Alone she worked this homestead land, and they all survived. There were times she said, she believed that they would not.

One very cold winter when her son William was five years old, he helped her cut down a cart load of poplar trees to take into town to sell as firewood and to buy salt, flour, sugar, and coal oil for the lamp. The cart was a two wheeled cart pulled by an ox.

On their way to Innisfree, then called Warwick, they were sitting on top of the loose load of poplars, wrapped in blankets, to stay warm during their journey, when suddenly one wheel of the cart hit an ice hole, the cart shook and they both fell off the top. Annie's leg was broken in two places. With her son's help, together they made a splint using some poplar pole sticks, and they tied the splint to her leg with binder twine; and then somehow reached the village of Innisfree.

Years later, when she was my grandmother, she was telling me how my father was growing up in the wilderness in northern Alberta, and I asked her if she was taken to a doctor in Innisfree (Warwick) when her leg was broken; she said, "Oh No". Her legs were both wrapped in gunny sacks for warmth. She had no stockings on, embarrassed, she would not go to see a doctor. She had been out in the woods cutting down trees. Her stockings she kept to wear to church.

Her son William sold the wood and with that money bought the supplies they needed as she waited in the cart. My grandmother never talked about being poor, nor did she ever consider herself to be poor. Money was not the issue. She was being thrifty. Money was not God given and did not matter. They had good health, a family together and fertile land to grow things on for her and for her family to survive on. That was her wealth. When they returned to their homestead cabin with their supplies, together they made a set of crutches for her and when springtime arrived, she was outdoors seeding and planting her garden.

My grandmother's daughter Julianna was thirteen years old when she married her man Paul Kotowich. Paul was seventeen years old. A man of knowledge, he was a writer of books and poems. My grandmother left her homestead to them to live on and to care for as their own when William was 8 years old. She then married again; a widower from Mundare. His name was Mike, who had five sons and a homestead in Mundare. Mundare was a country town with the homestead at one end of the town's road. Mike did not want grandmother's eight year old William, her only son, to live with them. He said that his five boys were enough to have with them in their two room cabin home. William continued to stay on the homestead with Julianna and Paul, missing his mother, and my grandmother was heartbroken that her only son was not allowed to be with her. She had no say. A wife had to obey her husband. Mike was not a strong man. His health was failing him which made him unable to do much work on the homestead, this left my grandmother alone with that task, while Mike's five sons were attending a Basilian Fathers St. Peter and St. Paul Catholic school in Mundare. My grandmother grew a garden and sold her vegetables to a "Howryluk General Store", and she cleaned and dressed geese, ducks and chickens to market there. She also did all the gardening for the priests. During those years as pioneers in Canada, there was no government pension, welfare, nor was there a food bank.

Though still a child on the homestead with Julianna and Paul, William worked like a grown man. He fenced, plowed, planted and seeded the land. He chopped wood and milked the cows. At one time when I was a little girl in school, my father told me that after he had started going to school, he learned some swear words and when the work on the homestead got to be a little bit much for him to handle and kept him from doing his school work; he used those swear words in conversation and Paul would always politely tell him what those words meant and that he was not to say them. William trapped and skinned beavers. He sold their pelts and with this money he put himself through high school in Innisfree. There were many times when Julianna and Paul took his pelts, sold them and spent the money drinking, and buying drinks for their friends till the money was gone.

At 18 years, and after his high school education, William left the homestead to work in a paper mill in Washugal in Washington State. But first he went to Mundare to see his mother. Mundare was celebrating "The Green Holiday" people had gathered from many miles around. There he noticed a young lady, the most beautiful girl he said he had ever seen, who was with her parents. Her long fair hair was rolled into a bun covering each ear. When he got brave enough to talk to her, he found she was a farm girl from a homestead near Hilliard, from a wealthy beef and dairy farm. Her name was Katharine.

He could not forget Katharine while he was working at the paper mill. He saved all his money and three years later he returned to marry the girl of his dreams. Katharine was five foot two, eyes of blue, with light brown hair and skin as white as snow. Her ancestors were French, Polish, Austrian and Ukrainian.

William was six feet tall, a handsome man of power and strength with dark hair, dark sparkling eyes, and his skin was the colour of a suntan summer and winter.

They were married June 20, 1926, in a little church in Mundare and there he became the district barber and owner of a liquor store and pool room.

William's mother was alone again. Her second husband Mike, had died and her five step sons were all grown up and away. She was happy to have her own son William with his bride Katharine near, after all her many previous sad and difficult years in Mundare without her only son. Remembering his father Michael to whom she promised that she would care for their only son and then, she found she was not allowed to.

Before Mike died he donated their homestead to the Basilian Fathers in Mundare, leaving their two room cottage with a large yard, a chicken house and a koomora. (the koomora was a one room log building. Perhaps the first home built on that homestead. In the museum today, 1993, the koomora is referred to as the summer kitchen). The yard and the buildings were left for my grandmother to live there till she dies. Today a museum occupies the yard. The koomora is situated in front of the museum and the two room cottage is situated at the back door of the museum.

Mike was completely blind and was not able to stand or walk for many years before he died. My grandmother alone nursed and cared for him for many years till his death; and she alone raised his five sons.

She continued to live in her two room log cabin next to the little church though she owned a big house in downtown Mundare. How she acquired that house is not known to me, though knowing my grandmother and knowing the love and

William and Katharine on their wedding day, June 20, 1926.

William and Katharine with William's mother and Katharine's mother and father.

1926 - William purchased this car to take his bride on a honeymoon.

The church William and Katharine were married in.

compassion she carried within her heart for her only son, I can see her saving every penny she could save to have something of value to give to her only son; and that would be a home, a home she was not able to give him during his growing into adult years.

Grandmother gave her son William and Katharine that big house as a wedding gift after William and Katharine were married. Katharine's parents gave William and Katharine one of their best milk cows. This too was a wedding gift.

After William and Katharine moved into the house in Mundare; William, my father, built a little barn on the lot behind the house for the cow, and the barn in later years became a garage for a car. During those years the 1920's and because of no refrigeration in homes in the cities and the towns, many people kept a cow to supply their families with milk, butter and cheese.

Katharine never liked William's mother, my grandmother, and never had a kind word to say about her or to her and she was not happy living in Mundare in that big house. William never interfered. He left them to sort out their own differences as he loved them both.

A first baby daughter Elizabeth was born May 4, 1927, and died six months later in her mother's arms from an injection a doctor gave her for colic.

In April of 1929, William sold everything he owned in Mundare and moved his wife and a second baby daughter, Anne, born June 14, 1928, onto forest acreage near Viking; where he built a log house with a kitchen and breakfast nook, dining and living rooms, two bedrooms and a loft. He chopped trees to clear this land for growing grain. With horses he pulled the tree stumps and the roots. And he built a stoneboat to gather the stones and the rocks into. The stoneboat was a flatdeck six feet long and four feet wide and had runners like a sleigh built of logs. It stood twelve inches above the ground. This was an exhausting and endless task, unimaginable today.

Using his skills, as the Great Depression deepened, William built a log horse barn, a cow shed, chicken coop, pigpen and shed, two granaries, and a garage for his Model T Ford, and farm machinery. He worked hard and he was happy. For the first time in his life he had a family, to him a family was everything. His work was done to perfection. Everything he did or built was done to last. With poplar poles and barbed wire, he fenced this quarter section and it is today as firm and as strong as it was then.

I was first born on this farm on April 9, 1930, Eugenia Irene. Mother told me that a Gypsy midwife delivered me. Ever since I can remember, I was "Daddy's Girl", and I was "Daddy's Helper" and I followed him everywhere and I helped him with all the work I could.

My father built a pumphouse around the well and the pump that drew water. A cream separator was kept there to separate the cream from the milk and to make butter and cheese. Our refrigeration system was a hole in the ground eight feet deep. We called it a lidonia. Every winter water was poured in. After the water froze, a layer of sawdust was put down. Then more water and more sawdust, in layers till the hole was frozen to about half full. The lid was made of wood, and lined with flour sacks filled with sawdust. When the lid was put down a layer of soil was spread over it. This kept our food cold in summer, and not frozen in the winter.

March 26, 1932, a miracle. A son Edward was born. He had fair hair and blue eyes and from then on, he was mother's and father's pride and joy; their first son.

The well on farm near Viking. Katharine holding 9 month old Anne.

William and Katharine on the farm

Katharine with her mother's sisters in front of the wedding present home.

Baby Elizabeth

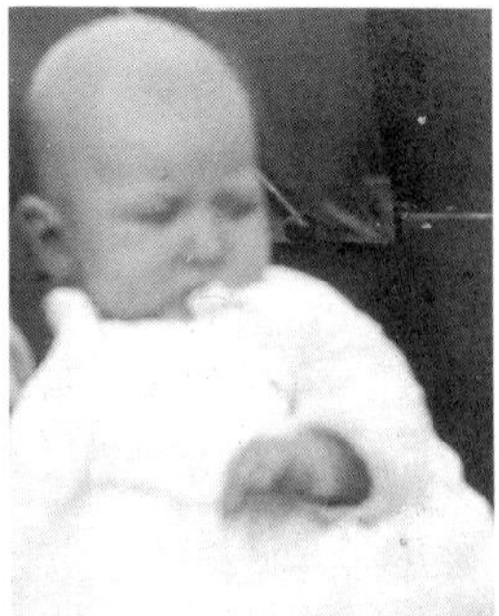

Two years later, Henry was born on April 28 of 1934. He was twelve pounds, very large, and had dark brown eyes and no hair; he was bald. Two sons and two daughters made life more complete for my father and mother. After seeding grain that spring, 1934, we all went to Mundare, to celebrate the "Green Holiday" in our Model T. Mundare was a small country town with dirt roads and boardwalks. A post office, general store, a hotel, barber shop, pool room and a liquor store. The main street was a wide dirt road and all the buildings were decorated with sweet smelling, spring budding black poplar branches. The black poplar trees were planted in barrels in a row right through town in the middle of the road. Horses, wagons, Model T's, carriages and tents were everywhere. The sound of music, the gala, the dancing started at dusk and went on till dawn, with everyone singing, dancing and talking.

In a field over an open fire, a young pig and a steer were being barbecued. On a long homemade wooden table, covered with white tablecloths embroidered and crocheted with lace, were preserves, fresh homemade breads, country churned butter, saskatoon pies and whipped cream, home-cooked foods of every variety and chokecherry wine. The "Green Holiday" was celebrated in Mundare every spring for a prosperous year ahead. Amongst all this feasting, the first ever built airplane landed, a biplane. It had two wings on each side. The pilot jumped out and joined the celebration, and then he took people who were brave enough to fly, up into the sky, while everyone else watched. He asked my father to go up flying with him and my father's reply was, "I'd rather keep my feet on the ground". Soon the food was gone, the toasting done, the music stopped. The "Green Holiday" celebration had come to an end. People hitched their wagons, gathered their tents and families and headed out; back to their homestead homes.

Mundare again was quiet - except for the mating call of a bullfrog.

Main Street Mundare, early 1900's

Chapter 2

Grandmother stood at her garden gate, waving a white handkerchief as we were driving away; towards Vegreville, with the sun setting, casting a golden haze over Mundare and reflecting off the church steeple. Only the echoes of the chatter, the music, and the laughter were still ringing in our ears.

Vegreville was a town of fancy shops; a tailor, one owned by Bill Zaruby, who was my mother's cousin, and who designed and made clothes for special occasions. There was a delicatessen, a bakery, a shoemaker, and a big hospital.

Ranfurly, a village of grain elevators could be seen from a distance. As we travelled on toward Ranfurly the setting sun was casting its light onto the gabled roof tops. The grain elevators, they were so high, they seemed to be reaching the sky. When a cloud drifted by, the roof tops were hidden in the clouds. Off the main highway to the left and over a railroad track was the Ranfurly post office and general store. A hotel with many rooms for overnight travellers, trappers and road gangs who were building the country roads, was proudly situated across from the post office; all brightly lit with gas lamps. There was a drinking parlour; gramophone music and dancing girls. The surrounding area was peppered with little houses, gardens and boardwalks and older people stumbling around growing their own gardens. These houses were all very small, they were built for the pioneers who left their homesteads to their kinsfolk while they retired to Ranfurly. Ranfurly is a neat and pretty little town - well kept.

Off the road to the right, travelling south towards Viking, around a small lake, and thirteen more miles along a country road and then home.

Oh country road, take me home

Past farms, up, down, and around duck ponds we see our pasture with our horses, colts and fillies at this, the north end, of our acres.

At the opposite end of our pasture, near the schoolhouse were the cows, lying close together chewing their cud. To see them there with their mouths going, it looked as though they were having an important meeting, or a discussion. A few others were lazily wandering about with their calves. Horses and cows stayed in separate sections of the pasture; they did not mix.

Separating the pasture from a field of tall green oats, we played hide-and-seek in and flattened father's oats, was Lornedale school. This school was built on a half acre of our land. It was a one room school house with one teacher to teach grades one to eight. An artesian well in the yard bubbled up cool clear water summer and winter, and formed a tiny stream drifting through our meadow;

There's a tree in the meadow
with a stream drifting by
and where'er I go
I'll always know
I'll love you till I die.

And I'll always remember the love in my father's eyes, and the care, and the energy he put into building our home and developing our land.

Driving past the school house, then by the tall green oats, we stop to open a gate and down a driveway of tall trees forming an arch above us and; then around a circular driveway of roses, we park in front of our house. My mother grew a rose garden and everywhere were pink and red roses, and dahlias, of many colors.

The saskatoon, chokecherry and pinecherry were bordering the trees separating the house from the barnyard. At the back of our house was a big pond, home for wild and domestic ducks, geese and frogs. When I was two years old I used to watch the little ducklings and the goslings swim and play. When they gathered near to where I was, Mother said I would pick them up out of the pond by their little necks to play with and sometimes brought them into the house; carrying one in each hand by their little necks.

A pathway through thick trees in front of the house was the horse barn and to the left of the horse barn was the cowshed.

A walkway from the house next to a huge wheat field was our summer kitchen. The summer kitchen had a wood stove, a wash stand and a basin, table and chairs, a china cabinet, a big double bed and some cots. Our hired men stayed in this summer kitchen while working on our farm and the nanny housekeeper had a room in the attic of our house. During the winter months our nanny, Lenora, May or Hazel Sutter shared the bed with Anne and me because the attic wasn't heated.

To the right off the main driveway was our pumphouse, from the pumphouse and back towards the country road was the garage for our 1932 Ford V8 Model 18 Ford Car and Hart-Parr tractor. The garage was completely surrounded with poplar trees except for one large, tall Alberta Spruce that stood tall and proud near our entrance gate.

From the yard to continue due North was a wide fenced walk and driveway to the oat field, to the schoolhouse, and to the pasture. Before the pasture gate behind the school yard was a frog pond and many trees with huge magpie nests hanging from their branches. There were crow's nests and many other birds. The orioles, red-winged blackbirds, meadow larks and robins nested in smaller trees away from the crows and magpies. After a rain fall, there were many large Orange-Top (Koozarie) mushrooms growing around this pond. We were out there picking buckets full of them after every rain fall. Mother cooked them in sour cream and green onions; and always they were a delicious treat.

Father loved the outdoors, all the animals,
the birds, the bees, all the flowers and the trees.
And he sang as he worked and he danced
to the music of any fiddle that played.

The aroma, and the perfume, of roses, the blossoms of saskatoon, chokecherry and pinecherry trees filled the air surrounding our house and yard, making a beautiful garden. With the silver willow shrub trees growing in hedges along the roadways, the open fields of tall golden grain waving in the soft summer wind like waves on an ocean and everywhere birds were singing, chirping and building their nests. The bumble bees were so big they waddled along the ground too big to fly. Whenever father left his work coveralls hanging outside on a hook overnight; in the morning he'd find wrens nesting in the pockets. The coveralls would be left hanging there till after the little wrens hatched and flew away. At the far end of the yard across from the garage was the chicken house. Between the chicken house and the horse barn was the pig pen and shed.

Mrs. Stewart was the midwife in our neighbourhood when Edward was born the morning of March 26, 1932. She was not able to come, and the only people in the neighbourhood who had a telephone, were the Charlie Bows. From Stewart's, father drove to Bows and called the doctor from Viking. When the doctor came he was drunk, but he found his way to the farm and delivered father's first son.

All our hired men; though they stayed in the guest house, they ate their meals in the house with us. The year Edward was born; the hired man, after working out in the field all day came to the house and with him brought his violin. After he had his supper, he sat on a wooden chair in the kitchen he was left handed, and he played that violin so beautifully I was mesmerized by this beautiful violin music. I do not remember his name though from the day I first heard him play, I too wanted to play a violin. Anne, Edward, Mother and I, would all be comfortably settled in father's lap listening to this beautiful music for hours before bedtime. Our nanny housekeeper, a young lady by the name of Tillie Karaym, later also sat on father's knee and gave him a big hug. For this act she got fired. Mother made sure of that:

Because, I don't know.
Because, she loved him.
Because, he was wonderful
Because, he was her Bill.

The next young lady hired was Maggie Kesylevich who kept her distance from father when she heard why Tillie was fired. During all our growing up years on the farm, every evening while our nanny was cleaning up after our supper and washing dishes we were always in father's lap, and mother too, with some of us on his shoulders while he was teaching us different songs, telling us stories, or telling us what he did all day out on the field; of how many Prairie chickens nests he found while plowing and of how many gopher holes he turned over with the plow. He also wanted to know everything that we all did during the day.

Father was also always concerned and interested in keeping up-to-date with what was going on in the whole wide world. Now in our fourth year of the Great Depression; perhaps why he chose to leave town life the year 1929, and to move onto a farm for a better life. We had no telephone and no newspapers were delivered to the farm. Early winter 1933 father brought home an R.C.A. Victor Radio, a floor model. After the violinist left for the winter, we all sat on the floor around this radio with our ears against that wooden box listening to a whole lot of talk, not understanding a word of it; though father did. He was interested in world politics. Richard Bedford Bennett, a Conservative, took over as Prime Minister August 7, 1930 from Liberal William Lyon Mackenzie King. Father was interested in changes, and if there were any changes for Canada.

Adam Nagelsky was the man hired to help with the farm work the following spring. Father praised this man many times; he said Adam was a good man. Our nanny housekeeper that year was a neighbour's daughter, Lenora Sutter.

Adam helped with plowing the fields, harrowing and seeding. He also helped with training young horses to pull a wagon, a sleigh, and other farm machinery. When the grain was ripe and ready for cutting, father cut the grain with a horse drawn machine that cut, bundled and tied the grain. Adam did the stooking and helped tend to all the animals. A stook was one bundle of grain set firmly into the stubble with the grain kernels facing upright; then three bundles on each side leaning against that one. A total of seven bundles of grain to one stook. Lenora the Nanny helped with milking cows, separating the milk, washing clothes on a wash board and tending to the children. Mother did all the cooking and baking; homemade bread, buns, pies and the best chocolate sour cream cake ever made. This was her own recipe. And the do-nuts she made were Oh-so good, some she filled with cooked and mashed beans sweetened and mixed with raisins, and she made all our clothes, coats, dresses, pants and underwear. Our underwear was made from flour and sugar sacks. Father bought flour sacks at a mill in Edmonton because they were cheap, 100% cotton and very strong. These sacks were also used for making tea towels, guest towels, table cloths, and bed sheets. Our bloomers, petticoats and briefs had the design, and the face of Robin Hood. A flannel material was bought for making diapers.

Father took our own homegrown wheat to a mill in Innisfree to be milled into flour for our baking needs. He did not buy flour. This mill in Innisfree also made our porridge; oatmeal, barley meal, cornmeal, cream of wheat; and this mill also made chop, for the horses, cows, pigs and chickens. Chop was all the grains chopped together; this was fed to the pigs and the chickens mixed with skimmed milk. Horses and cows were served the chopped grains dry.

Autumn harvest time was a gathering of all the grain growers who had threshing to do. Men with their machines, horses and wagons, moved from farm to farm threshing each other's grain and hauling it to the grain elevators in Ranfurly. The threshing machine blew the straw onto one big pile while wagons with teams of horses collected the grain coming down a chute from the machine. At each farm, wherever the threshing was being done, all the woman came; they brought all their children and did all the cooking for the threshing gang. My sister Anne and I were sent out one day with the mid-day lunch for the threshing gang. Paul Kryzanowski was the owner of the threshing machine and he sat high on his tractor watching the bundles being thrown into the machine, the straw being blown out, and the grain collected. The machine was very noisy. We came up to the tractor and were yelling at Paul Kryzanowski "Your lunch is here". Sitting high above us, he did not see us, nor did he hear us. I carried a big silver tea kettle full of hot coffee to the field. The coffee was made of chicory and oven roasted whole wheat, oats and barley mixed with some coffee. After a few calls, and being ignored, I picked up the teakettle of hot coffee high above my head and I poured some out of the spout into his shoe. This made him shake a leg, look, and stop the machines. The gang then gathered around and had lunch.

Every late autumn before the snow fell father always worked with teams of horses building country roads. This was to work off the country taxes due every year. Covered with soil, tired and hungry, he'd return and he'd sing;

Katie beautiful Katie
You're the only girl that I adore
When the moon shines
Over the cow shed
I'll be here at the kitchen door

When autumn leaves had fallen down, the red and yellow, gold and brown, when the whole wide world was calm and still, waiting for the snow to fall; father took a last wagon load of grain to Innisfree to be milled into flour, porridge, and chop to be our supply for the coming winter months.

Chapter 3

Slowly, he walked through the town to the big brick high school and as he stood there reminiscing of the days of his youth, the homestead, of his beaver pelt, and of his high school days; completely lost in thought. A wagon went rumbling by at full speed, going towards the mill, leaving a cloud of dust behind them. It was Julianna and Paul whom he had not seen since his wedding day in June 1926. They were on their way to the Mill for their supply of flour, porridge and chop for the winter.

He walked back to the mill; his order was ready to go. He loaded his wagon, brought his team of horses around from the stables, preparing to leave when Julianna and Paul approached him and persuaded him to wait awhile and to join them, in the Ranfurly hotel for a few drinks. He had to go to Ranfurly, to get onto the country road going south. That road would take him home, and he agreed to stop awhile.

The drinking parlour was full of local retired old men, and young farmers of the north; all were chattering as loudly as they could to be heard over and above the loud gramophone music. Julianna was the only woman there, besides the dancing girls.

My father, his sister Julianna, and his brother-in-law Paul whom he had lived with for most of his young years talked and they drank, and they drank and talked some more. And they laughed and they cried as the hours slipped quickly by. Father realized he must be on his way; he said, Katie will be waiting with his supper. He tried to get up and fell back into his chair and he passed out; not being used to drinking liquor; he was drunk. Julianna and Paul carried him out to his wagon and placed him amongst all his sacks of flour, porridge and chop; then fastened the driving reins from the horses, firmly to the driving line hold in the wagon, then they slapped Queenie and Pete on their rear ends telling the team of horses to take him home. Pete was a Percheron dapple gray horse, with a white mane and tail and Queenie was a Belgian light bay mare with a creamy-white mane, tail, and the long hairs on her ankles, were also a creamy-white, away they went; clippity clopping around the small lake, then south along the country road; they were on their way, to take him home.

The clippity clopping sound of horses hoofs on the country road, the cool air, and the rocking and rolling of the wagon wheels made father realize he was on his way. He sat up, and looking towards his team of horses, he could see no driver; so he sang to the horses;

Where'm I goin'? I don't know
Where am I headin'? I ain't certain
All I know is I am on my way.

He then fell back amongst the sacks and fell asleep. From the sound of his voice, the horses knew he was with them and happily carried on clippity clopping along - with their harness chains clanging.

Several miles along in the dark of the night; the sounds of timber wolves howling disturbed father's sleep. As he listened, he heard Pete and Queenie still clippity clopping along at a faster pace. He then looked into the trees on the side of the road and could see eyes shining; the wolf pack were following the wagon. Again he sang louder. To the horses; he asked;

When will I be there   I don't know
When will I get there?   I ain't certain
All I know is I am on my way.

His head was spinning, his whole world was turning upside down, he felt sick. Leaning over the side of the wagon, he threw up some of the alcohol he had drunk and fell back into the wagon. His head felt as big as the sky above and his eyes twinkled like the stars. This night was dark, with no moon to light their way and it was getting very cold. He was not sure where he was, all he knew was that he was on his way. As he lay there in the wagon looking up into the sky; at a distance, he heard farm dogs barking. Queenie and Pete were trotting along at full speed now moving quickly, the wagon was bouncing, rocking, rolling, and the wheels were turning faster He sat up, looked out and he saw they were passing the Lornedale school yard; soon they would be home. And he got to thinking; what lies ahead.

Where am I goin'? I don't know
When will I be there? I ain't certain
What will I get? I ain't equipped to say
But who gives a damn, we're on our way.

Before he could decide what he must do, Queenie and Pete turned and stopped, at the main gate. And there to greet them was Hero, the black Spaniel and Rex the Collie, wagging their tails, jumping up and down excited and happy to see them. They had been waiting at this gate, for their return.

Slowly father slid off the back of the wagon; holding on to the wagon, then to Queenie's harness as he made his way to the gate to open it; he fell a few times before he was able to swing the gate open. When the gate was opened and the horses and wagon went through, "Whoa!" he said; the horses stopped and waited while he closed the gate. Grabbing onto the back of the wagon, "Giddy up," he said to the horses and away they went as he walked along holding on to the wagon. Rex and Hero were running along in front of the team as their pilot.

In the house, he could see a light in the window and he knew mother was waiting. He stopped the horses by the pump house door where the sacks of flour, porridge and chop would be unloaded. The sacks were all neatly covered with a canvas tarp, so he decided he'd leave them there till sun up. He unhitched the horses from the wagon, leaving the wagon and the load by the pump house, slowly he walked the horses to the barn where fresh sweet hay was waiting for them in their stalls. There he slipped the harnesses off Pete and Queenie, he then closed the barn door and started his long wobbly walk towards the house. He came stumbling and falling along the pathway with Hero and Rex jumping up and down beside him and tripping him. Mother stood at the kitchen door; she was worried, she was angry, and she was crying. When she saw him stumble and fall, she ran out to help him in. She was so happy to see him that she called him Billy. (Vasailkoo mei)

Father did not have his supper when he came into the house, but went right to bed. A few hours later when he was wakening, the sun was high in the sky the morning sun had warmed the land, the whole world again seemed warm, at peace and safe. The first words he said to Mother were; "Katie! this is the first, and last time."

And it was; he never again got drunk. 'Twas the horses; yes, here's to the horses, the reliable, tried, tested and true. They brought father home.

Horses are superhuman beings. They have a special devotion to humanity, they know what is expected of them and they do their utmost with energy to please their master. In return, all they expect is kindness, appreciation; a brushing, a handful of oats, a carrot, an apple and talk.

During the developing years, what could our pioneers have done without horses clearing this land, farming, travelling, delivering milk and more? Building Canada was done by a horse. Horses are very special

Forty years later, during the year 1973, Father's first born son Edward had a cattle ranch in the Cariboo of British Columbia, besides his teaching profession. Winter was approaching; it was round-up time. All the cattle were brought in, to corrals for the winter, to be fed. After most of the cattle were in, Edward went out alone, on horseback to look for the strays, searching the vast hills, meadows and trees of the Cariboo. He rode for hours through thick trees, over hills, and the meadows searching for the cows and the calves who had separated from the herd; when he realized it was dark; night had come and he was lost. And here's what he said;

I talk to the trees,
but they don't listen to me.
I talk to the stars,
but they never hear me.

The breeze hasn't time
To stop and hear what I say;
I talk to them all in vain.

But suddenly my words
reach someone else's ear;
Touch someone else's heartstrings too.

I tell you my dreams
And while you're listening to me,
I suddenly see them come true.

His horse reared, he would not go in the direction Edward tried forcing him to go in. Edward tried and he tried by talking to the horse and by slightly kicking him in the ribs as he sat in his saddle. It was getting very late, the ground was frozen with a sprinkling of snow everywhere. He could not see if there was quicksand ahead, and could not understand why his horse would not go in the direction he thought was home. His horse was a Morgan horse named Mac. Edward was tired, hungry and much too cold to argue with a horse, so he hung the driving rein over the saddle horn and he said;
"Where'er thou goest, I will go." He then lay close to the horse, dodging low branches of trees and covering his hands with the horse's mane, to keep them warm, leaving happy

trails to Mac; his thoughts slipping into song;

I can see us on an April night
sipping brandy underneath the stars;

Reading poems in the candle light
To the strumming of guitars.

I will tell you all the books I've read;
And the way I met the King of France.
Then I'll send the servants off to bed
And I'll ask you for a dance.

Unnoticed the hours had been slipping by, when suddenly, he realized they reached familiar ground. His ranch house was in sight. Morgan Mac brought him home from the hills.

For fifteen years Edward ranched the Cariboo wilderness, and whenever he went out into the hills looking for cattle, hunting, or just exploring the vast Cariboo Country; he went on horseback, or with a team of horses and a sleigh leaving his snowmobiles and skidoos behind.

Horses are very special and necessary in forbidden, unfamiliar places. Always horses will get you home if you let them.

Edward with his covered wagon team in the Cariboo.

1927 - Father's sister Paulina, his mother, and Katharine

Father with sisters Juliana and Maria

1933 - Father holding Edward, Anne, mother and Eugenia Irene in our yard

Chapter 4

Winter nights were crisp and cold, the cracking sounds of someone walking could be heard from far away. On a moonlit night, the moon shone a big silver bright; or Northern lights, all the colours of the rainbow, danced in the sky the whole night through.

The air was fresh and clean.

Sometimes in the springtime, there would be a mirage. We could see far away places. It was as if the world was one big rubber ball, with some air punched out, and the cities and the towns were in that hollow. And we could see them.

Mrs. Vern Gilpin, a Registered Nurse, was our neighbour to the south. Every school day, she took her daughters, Meryl and Myra to Lornedale School in a one horse sleigh in the winter time, and a one horse buggie when the snow was all gone. Everyday, on her way home she would stop in at our farm to check on mother. She knew a baby was due. April 28, 1934 at 6:00 a.m., a Saturday morning, Mrs. Gilpin was there to deliver Henry, son #2. The doctor from Viking arrived much later in the day to make sure the delivery was OK.

Father made sure a nanny-housekeeper was always there to help mother with the housework and the children. Nancy Matchuk was our Nanny when Henry was born, though she left two months later as she was planning to be married. Her young brother Bill was hired to help with the farm work, and Jenny Orenchuk, mother's youngest sister was hired as Nanny-housekeeper.

Bill Matchuk was afraid of horses and would not go near a cow. He was also afraid to stay in the guest house alone. He brought all his quilts, blankets and pillows, and he slept on the floor, in our dining-living room. One night father found him crawling along the floor with all his covers over him, talking in his sleep, saying whoa Topsy whoa. (Topsy was a very tall mare and full of spunk.) He was afraid of her.

Father kept Bill Matchuk employed and he paid him well because Bill was Paul Kryzanowski's brother-in-law and Paul Kryzanowski was the owner of the community threshing machine. Bill and Jenny often used to sneak away after the supper hours to be alone, and to go for long walks in the moonlight.

One late August evening when Bill and Jenny were away, a dark storm cloud, with a strong wind blowing was seen heading in our direction covering the setting sun. Mother went with father, out into the field with a team of horses and a hayrack to pick up the dry piles of hay ready for storing in the barn loft before the wind would blow them all over the field. Mother drove the horses from pile to pile while father pitched the hay into the hayrack.

Anne, Edward, Henry and I, were taken to our summer kitchen before they left. From the summer kitchen we could see where our father and mother were and would not be afraid to be left alone.

The wind came howling, rattling the windows and the door, roof shingles were flying everywhere in the yard. Henry needed changing, and would not stop crying. He was a breast fed baby, and he was hungry. Anne and I lit a coal oil lamp, and did our very best at changing the baby's diaper. The diaper was fastened together with a big safety pin. We were not sure how to fasten the diaper without sticking the pin into the baby. We then heated some milk in a little pot on the stove and while Anne held the baby in her arms holding a nipple in his mouth; I poured warm milk out of the pot into a little pitcher, from the pitcher into that nipple as she held it. As I poured milk into that nipple the baby squirmed, tossed and turned, legs and arms going in all directions. Before he got enough to drink he got a complete milk bath, and so did Anne. Edward stood by and watched.

Anne was six years old; Edward two, I was four, and Henry four months.

The wind blew stronger as the hours passed. We were very frightened being home alone. It was getting dark and very late and we could not see father, mother and the hayrack with the horses.

Before long, mother and father rushed home with just the team of horses leaving the hayrack piled high with hay out in the field; because they said the horses were going wild and could not be controlled. The horses knew what the storm was all about, and wanted to get home. After putting the horses safely into the barn, father and mother came running to the summer kitchen. They wrapped Anne and me into quilts and pillows and sent us under the bed while they held Edward and Henry in their arms. The tornado spun, turned, tipped and tumbled everything in our yard. The hayrack that was left out on the field was picked up and carried into the next field and set down with the hay still in it.

Farm buildings near Birch Lake; a lake north east of our place, were completely destroyed. A barn was picked up off its foundation, carried away and was never found; leaving the horses standing and still tied to their stalls.

Roofs were torn off buildings, and some buildings that were picked up and carried high up into the sky; when dropped, broke into tiny pieces.

Our nanny and hired man did not come home that night. They stayed the night with Sophie and Paul Kryzanowski because of the storm. Father gave them both a serious talking to when they returned the next day. Jenny in tears, wanted to go home. Father and Mother together drove Jenny back to Hilliard to her parents home and brought her older sister Agnes. Our hired man Bill was left to look after us, the children, while mother and father were away, he wanted to surprise them by having our supper all ready for us before they returned, he made a custard pudding, using milk and clothes starch instead of cornstarch. Because of the clothes starch this pudding boiled out of the pot all over the stove and by the time he scraped and cleaned the stove, mother and father had returned with Agnes. They were happy to know the custard pudding had run all over the stove before we ate it or we all would have been very ill. Agnes stayed only a few months. May Sutter was hired as our nanny-housekeeper for that year. Lenora, May and Hazel Sutter who were our nannies from time to time, were the daughters of Charlie and Cora Sutter who were our neighbours to the West, on the next section of land. They also are family to the next generation - the famous young men the Sutter hockey players.

Some years later, when Bill Matchuk grew up, he married a girl named Katie, and together they owned and operated the biggest and the best motel in Radium Hot Springs. My father and mother often stayed in that motel, and so did I and the rest of our family who used to go to the hot springs.

Mother's sister, Jenny, lived her life in Kitchener Ontario common-law with a married man from Saskatchewan, who was much older than she was and her sister, Agnes, married a Service Man, a United States Air Force Officer, and lived most of her life an American in Buffalo, New York.

Thanksgiving was celebrated every year, the month of October in Lornedale School. All our neighbours brought to the school, sheaves of grain, corn, hay and some of the vegetables they grew in their gardens. Preserves and hand-crafts; knitted sweaters, socks and mittens, embroidered and crochet tablecloths, pillow slips, tea towels and much more. The school was packed with beautiful things, and the smell of grain and vegetables filled the air. Every year at the school for Thanksgiving, there was a concert or a stage show, and a Thanksgiving Church Service. The food, and the clothes that were brought to the school, were left there for the church people to give to the people who might be in need wherever they may be.

Cabbages mother grew in our garden one year were so very big that one cabbage could not fit into the stoneboat. Father had to bring them in from the garden in the hayrack. The cows, the chickens, the pigs and all of us had cabbage to last and last.

Mother made barrels full of sauerkraut. The barrels were left outside for the winter and froze solid with a big ice bump in the middle. Whenever she wanted some sauerkraut brought into the kitchen for cooking; father chopped it out with an axe.

In some areas of Southern Alberta, there were periods of draughts in the early 1930's. People used to pass by our farm, travelling with horses and wagons. All their possessions were in their wagons with them just like the early pioneers. They were heading for Grand Prairie and the Peace River. Often they stopped at our farm to water their horses and to fill their canteens. Father always gave them milk, butter, eggs and chickens for their journey. One traveller whose name was

Steve Danchuk, had a Maytag wringer-washer he wanted to sell. Because father did not have enough cash on the farm that day to buy the wringer-washer; the man accepted two of our best milk cows as well as the cash father had. Mother did not want the Maytag washing machine. She said it was much too expensive, and because Steve Danchuk was father's brother-in-law mother wanted nothing from him.

Steve was a big man with a loud voice, red curly hair, a red mustache and deep blue convincing eyes. He was always on the go; travelling from town to town buying and selling whatever he could. He was married to my father's oldest sister Maria and they were on their way from the south to settle in the Peace River. After they were settled there Maria stayed on in the Peace River, and alone raised twelve children. Steve continued travelling, buying and selling anything and everything, his whole lifetime through. And obviously he did get home to the Peace River from time to time to keep Maria barefoot in the summer and expecting babies in the winter time.

Father bought the Maytag washing machine for Mother; he did not mind the price. The machine had a little gas-oil motor. When it started with a kick start it sputtered and puffed smoke, sounding loud like a threshing machine. It had a lever on the right side of the tub to control the washing speeds, and to stop the agitation before turning on the wringer. It was kept in our pump house because of the noise and the smoke. This Maytag wringer-washer kept on washing all our clothes non-stop, from the early 1930's to the early 1950's; a machine, father traded two cows and some cash for, from a travelling caravan pioneer.

Chapter 5

December was Christmas; always white merry and bright. People singing, children playing, jingle bells a ringing. Happy smiling faces everywhere preparing for the holy-day, Christmas.

On our wheat field next to the guest house was a man-made ice rink. My father made it with some help from the neighbours men. They were bringing buckets of water from the well by hand. In the guest house, a fire was burning in the old country stove with a kettle of hot chocolate sitting on the warm end of the stove over the hot water reservoir. On the table were stacks of hot chocolate mugs, a variety of sandwiches and cookies. Mother made the hot chocolate of cocoa, hot milk, sugar, vanilla and cream.

Neighbours and their children came to play on this ice rink. All the children from Lornedale School were there and so was the teacher. They played hockey into the late hours of the night, by the light of the silvery moon. My father was the School Trustee and he looked after all of the needs of the school, the teacher and the children. He fastened jingle bells to his horses' harnesses and he took everyone who came for sleigh rides over the fields of snow in search of Christmas trees for the school and for our homes. He set the hayrack onto runners with hay and straw in it for us to sit on and to keep warm in, and away we'd all go singing Christmas carols all the way. The tree branches and all of the shrub trees were layered thick with frost. There were many white rabbits hopping around through the woods, then would hide in the thick frost laden shrubs as we went by singing with jingle bells ringing . A grouse or a prairie chicken would rustle up from hiding then scuffle along the snow. This was a truly beautiful silver-white winter wonderland, fresh clean and clear.

During the day when the temperature warmed, from all the buildings huge icicles hung and when the sun shone on them they sparkled all beautiful bright colors like diamonds. They hung from the roof tops to the snow below.

December was the month for Christmas planning and shopping. Every year in December, father went by sleigh and a team of horses to Ranfurly where he left the horses in a nice warm stable and from there he took the train to Edmonton to buy gifts and things. The train always stopped in Mundare going to Edmonton, and on its return. My Grandmother was always there at the train station, waiting to greet her son, and on father's return from Edmonton, Grandmother always had a Christmas box of two to four of her home raised geese dressed and ready for our oven for our Christmas dinner. How she knew when her son, my father, would be passing through Mundare is unknown to me, with no telephone in homes, she was not taught, and did not know how to read or write - (though letters she received, she immediately took them to someone of knowledge to read to her). She could have been walking out to that train station every day, waiting and watching if her son was on it. I am quite sure that that is what she did.

In Edmonton, there was a woolen mill where father took all of his worn out woolen winter long johns to be made into blankets. These blankets were beautifully made, soft and warm. I still have a pale blue, wool blanket trimmed with a wide blue ribbon made from my father's worn out country winter long johns from the 1930s.

Christmas on the farm was a candle lit tree, decorated with crepe paper roses mother made, and crepe paper streamers and ribbons. The candles were real candles set into a little cup with a clip to fasten them to the tree branches. On our dining room table was a layer of fresh sweet smelling hay covered with an embroidered white tablecloth trimmed with crochet lace that my mother made. In one corner of the living-dining room was a pile of fresh hay with sheaves of corn and grain in the very corner and on either side of the hay. Crepe paper steamers and ribbon hung from corner to corner of the room and a paper bell hung in the middle from the ceiling above the dining room table. The fresh sweet smell of hay, corn and grain all through the house and the Christmas tree, is what made our Christmas so perfect and real.

Whitefish was delivered to all the homes in the back of an open sleigh drawn by a team of horses. This fish was fresh water fish and was frozen solid and tossed around like firewood. Whitefish was an important food for Christmas; some of it was made into a fish soup, and some was baked in the oven with dressing. During the month of December, Mother was busy cooking and baking for Christmas. She made three different kinds of cabbage rolls; these were, buckwheat, rice and ground beef, and rice with butter and onions. Perogys with sauerkraut and mashed potato, bacon with potato and onion, dry cottage cheese with egg yolk, and perogys with prune-plums. She made and baked potato garlic rolls. Potato pancakes with sour cream. A variety of donuts, cookies, and a Christmas cake. Apple pies, saskatoon pies and (Kootia). Kootia was whole wheat kernels soaked in water overnight, then simmered on low heat all the next day. When the kernels were tender she added honey and poppy seeds.

The poppy seeds were scalded first in boiling water then drained to almost dry, then she pounded the poppy seeds with a wooden pounder into a white milk texture before she added them to the wheat and honey. This Kootia was served in very small portions to family and guests before the Christmas Eve meatless banquet of twelve different dishes. These were:

1. fish soup
2. baked whitefish with dressing
3. buckwheat cabbage rolls
4. rice, butter and onion cabbage rolls
5. cottage cheese egg perogys
6. sauerkraut perogys
7. potato and onion perogys
8. potato pancakes with sour cream
9. hardboiled eggs
10. cheese
11. garlic potato rolls
12. (Ruegalki) sweet potatoes,

and all home grown vegetables off our garden: cabbage, turnip, carrots, parsnip, beets, some boiled and some pickled, dill pickles, mustard pickles, canned corn and baked potatoes with butter and sour cream. Christmas cake, plum perogys, donuts, cookies and pies with whipped cream were served after the main meals.

Ruegalki were a sweet potato we grew in our garden; they were very small, yellow in colour and shaped like a banana.

Our Christmas day dinner was roast goose from my Grandmother. Boxing day through New Years was roast pork and turkey with all the trimmings and homemade sausages (Kolbasa) and (sealcheeson) which was a headcheese my father and mother made. Homemade sealcheeson was made from home raised beef and pork. It was tongue, ears, feet, kidney, heart, boiled, spiced, and poured with the liquid into a prepared scraped, cleaned stomach of the animal and sealed. A round cutting board was put on top with a big rock on top of the cutting board to flatten the shape and to make it easy for slicing. Sealcheeson jelled solid when cooled and was easy to slice and delicious to eat.

All our bacon was home cured and it was from our own home raised pigs. It was cut into 6 inch squares, salted for a time then slightly baked in the oven; when cooled the bacon was put outside to freeze. After the bacon froze solid, it was stored in our refrigeration system, the ice hole (lidonia), in crockery containers.

I was three years old when father got me a little violin for Christmas. I was so very happy to get a violin, I started to learn to play it right away. This same Christmas Anne received an Orphan Annie doll; the doll was as big as she was. Edward and Henry got a sleigh to play with in the snow and a little red wagon for the summer. Mother always got a new dress to wear to the Lornedale school dance parties, basket socials and pie socials. From his trips to Edmonton, father always brought for all of us new shoes, new clothes and oranges, apples, candy and nuts for Christmas. Every New Year's Day, was spent driving in an open sleigh with jingle bells fastened to the horses harness. We went from neighbour to neighbour with our pockets full of kernels of grain. At their door we would sing a Christmas carol, and when their door opened, we threw kernels of grain into their homes. As we threw the grain, we would say

Сійся родися
Жито пшениця
у сяка пашниця
щоби мой Новий
рік був ліпший як
в торік

At her age of 91 years, this is my mother, Katharine's, handwriting.

This was a New Year's Day custom; to wish all fellow grain growers a prosperous and happy New Year: We would then be invited in for treats. All the neighbours were always very generous with treats. It was more like having a complete meal.

All the windows in our house were frost covered with beautiful designs of snowflakes and of leaves during the winter months. Looking at them, we could not find two snow flakes or two leaves that looked the same; they were all different and painted by Jack Frost. They were beautiful and amazing. Inside the house, around the windows and the doors was a thick layer of frost, making a border around the windows that looked like frozen sponge around the frost-work. The world outside with its thick frost covered branches on all the trees and shrubs, made our Christmases all Jack Frost decorated, and sparking silver-white.

The walls inside the cow shed were all white, thick, warm and cozy with frost. It was like being in an igloo when we went in to milk the cows. The frost on the walls sparkled all colours like diamonds from the light of the lanterns we carried. The cows and the calves that had to stay outdoors for the winter dug huge holes into the straw piles, big enough for them and their calves to get into. When they were hidden out of sight, they covered the opening entrance with straw to protect themselves from wolves and the cold. Wolves would sneak into the straw caves if they could, to keep warm. Father went out to the straw piles every morning to dig out the entry so that the cows and the calves hiding there would not suffocate. The very cold weather lasted from November through March and April. Often our noses, toes and fingers froze from playing outdoors. Mother thawed our frozen parts by having us hold a snowball over our nose and put our feet into a bucket of snow to thaw our toes. If we threw a dipper of water out the door, it would become a chunk of ice before it hit the ground.

In the winter, in sixty below zero weather through snow drifts higher than the fences, father made gunny sack nose guards for his horses, to protect their lungs from freezing and he put blinders on their halters to protect their eyes from the cold. He then hitched them to a sleigh and away he'd go, fourteen miles into town to Viking to sell cream, butter, cheese and eggs, and to pick up farm supplies. His eyebrows, eyelashes and mustache would be white with frost. Icicles formed with every breath he took, yet he never quit.

All winter long our clothes were hung outside on a clothesline to dry. The clothes would be taken out of the clothes basket real quick and before they could be hung out, they would freeze in all shapes and folds, and had to be left hanging there in funny shapes and folds. If we tried to straighten the folds out; the clothes would snap and break like an icicle.

I was five years old when I started school in September 1935. I wanted to go to school to be with my older sister Anne. She had already been in school for one whole year and I missed her. The teacher, Mr. Earl Hardy was a music teacher and was happy to have me start school; immediately he taught me to read music notes. Earl Hardy was from Viking; his father owned and operated the Red and White Store in Viking It was a general store, and is still there. During the months Earl Hardy taught school, he boarded at Vern Gilpins' farm. In the winter time he would whisk past our place on a set of skis behind a galloping horse, holding onto the reins. There was a small barn on the school yard for horses to be kept in during school days. Some students came to school from four miles away, some came on horseback and some with horse and sleigh.

When Christmas came that year 1935, I played my first violin solo at the Christmas concert. My Mother made me a beautiful robin's egg blue satin dress with a full gathered skirt, big puffed sleeves, and with a big bow tied at the back. I loved that dress, and I played;

Good morning to you
Good morning to you
Whatever the weather
We'll make it together
At work and in play
A beautiful day.

My first finger on my right hand, holding the violin bow was wrapped with a thick bandage; I had cut it to the bone with a razor blade helping to make Christmas decorations for the school. Lornedale school was decorated every year for Christmas with streamers, bells, ribbons and a tall Christmas tree stood near the stage. And every year for Christmas people came from miles around to see and to hear our Christmas concerts. There were gifts under the tree for all the students and a Santa Claus to hand them out.

Chapter 6

Early one morning
Just as the sun was rising

Jack Frost was gone and took with him all of the pretty pictures he had painted on our window panes; the sun came peeping through, warming our house. Robins were hopping in the yard looking for food. Our cows came out from their refuge in the straw piles, and were basking in the warm sun. Tiny streams were running along the roads down to the little duck ponds, and into our meadow. Green buds soon showed on branches of shrubs and trees, and blossoms were waiting to appear on the berry and cherry trees.

In the pig pen twelve pink piglets were suckling their mother, the Yorkshire sow. Hens clucked in the hen house making their nests. My mother put as many as ten duck eggs under some of the hens to hatch ducklings, and twelve to fourteen chicken eggs under some of the hens to hatch chicks. There were now calves, some colts and fillies being born.

Springtime.

Plowing, disking, seeding, harrowing and planting the garden. Father had left small town Mundare in 1929, because times were changing drastically. He had been managing the Silver Spray Brewery's retail outlet since 1925 which brought a good income, but in 1928, the government of Alberta took over the agency and paid for the stock of Silver Spray beer, leaving him without an income. He had graduated in 1921 from a barbers' college in Edmonton. He knew that barbering and the pool hall alone would be slow if hard times came, so he sold his barber shop and pool room business in Mundare after Anne was born on June 14, 1928, and bought 160 acres north of Viking. He moved his wife and baby daughter out to the farm in the spring of 1929 and he rented an adjoining quarter section and developed this parcel of 320 acres into a productive and park-like setting.

Hard times were everywhere during the 1930's and with foresight, father went into mixed farming - grain, cattle, hogs, chickens, ducks and turkeys. He said he was lucky that every year the grain grew tall and healthy. No chemicals were used, and no spraying was ever done.

With only a few hours of sleep each night, there was always much work to do.

Every spring, father, mother and our hired help plucked, cleaned and dressed 10 month old and year old grain and milk-fed poultry for the market. The birds sold for 15 cents. Ducks and turkeys were 17 cents each. A 200-pound select hog, chop and milk fed, sold for $4.00 and $4.50 each. A two-year-old shorthorn steer sold for $16.00 and $18.00 each. Buyers from Winnipeg came to our farm. Some buyers would offer as low as $15.00 for a steer while another buyer would offer $18.00. Father had to sell the livestock at these terribly low prices to have money to live on, but every year he managed; he kept us all well fed, clothed, happy and healthy. And he had money in the bank. Farmers today would never believe these Great Depression prices. Eggs were 8 cents per dozen, butter 10 cents per pound. A five gallon can of sweet thick cream sold for $1.50 per can. Clean, threshed, harvested oats sold at 6 cents per bushel and clean wheat at 18 cents per bushel. A bushel is a dry measure of 8 gallons (32 litres). These were the lowest prices ever.

Every spring father greased the wheel joints of his farm machinery before the spring work could begin. In 1936, Henry was two and Edward was four; they were both out with father, and saw him grease the machines. After father hitched his four horses to a disk and went out onto the field Edward and Henry went into the barn, took down the axle grease and not only did they grease the wheels of the other machines; they smeared the seat, the hand levers, the foot rest, the tongues the horses are hitched to and their faces, clothes and hair were black with thick sticky axle grease. They knew they were up to no good; but they were learning farming. For the rest of that day they splashed around outside in a tub of warm water and suds using mother's homemade lye soap.

Every weekend we all got a tub bath, from the youngest to the oldest in the same water. Father brought into our kitchen a big round metal tub and mother poured hot water into it from the stove reservoir, and that evening after supper, we scrubbed and then were sent to bed dressed in clean pajamas with mother's and father's hugs and kisses. Mother rolled my sister Anne's hair and mine in rags and when our hair was dry we looked like Shirley Temple Dolls, all in ringlets. My father cut and styled our hair, and all our neighbours' hair and that of their children, with his professional barber equipment. But he had to go to a barber for his haircuts.

At Lornedale School, the students from grade one to grade eight at recess played ANTI-I-OVER. Teams were chosen to line up on opposite sides of the school with a captain on each side to watch a rubber ball being tossed over the roof of the school to the team on the other side. The child tossing the ball would call out "ANTI-I-OVER" to warn the team on the other side that the ball was on its way. The side that caught the ball more times won. I was the youngest child, and could not throw the ball so Arthur Frances, a student in his final year at the school gave me piggyback rides during recess. Great fun! Then my sister Anne told mother about my piggyback rides and mother put a stop to them. Even today I believe it was because she did not want the others to see the Robin Hood face on the bloomers she made for me from Robin Hood flour sacks.

During the 1930's children respected each other, their parents and their teacher. Girls were young ladies in dresses, buttons and bows. Boys were young men, with their hair cut short; and they did not wear earrings. They held a high regard for all adults and grew up to be caring, trusting, responsible and hardworking. During the Great Depression, parents were in control and responsible in watching their children grow. There was no sex education in the schools: Mother Nature took over the sex education when young adults were ready for marriage. All the children I knew and grew up with had both, their father and their mother to be with to guard and to guide them through their growing years into adulthood. Most of todays children (our modern 1990s) are not allowed to be raised by their fathers; if a father spends more time with his son or daughter than he is legally given the right to, he is picked up by the police and put in chains and handcuffs and hauled into jail.

Convicted baby abductor Paul David McIsaac is banned from Vancouver Island, Judge Peter Doherty ruled Wednesday.

Doherty handed McIsaac a conditional one-year jail term plus three years probation for abducting his infant son on January 16, 1996.

According to police, McIsaac picked up the baby, Ross Smith, from his mother for a court-ordered visit. The father failed to return the baby at the appointed time.

The judge handed McIsaac a one-year conditional sentence and ordered him to leave Vancouver Island within 24 hours.

McIsaac is the baby's father, and was married to Smith briefly several months before he abducted baby Ross more than a year ago.

article form the Comox Valley Record, January 31,1997

During our modern 1990's sex education is being taught in schools, to the very young. And our very young are having sex and are having children at the ages of 12 and 14 years. Young mothers today have four to six and more children and each child has a different father. And legally, these young mothers alone are given the right, and are trusted, to be responsible to raise and to guide that child into adulthood. With a boy friend present. These young mothers are demanding money from the government, and from each child's father for them all to live on. Do they care: or do they know or realize that all the money in this whole wide world could never ever replace the love, the gentleness, kindness and strength and the guidance a father gives his child through his developing years. And the father teaches his child to be a caring, respectable, and a responsible human being. If our modern 1990s lifestyle continues into the future of the next generation; I can see the next 60 years, the years 2050s; fathers will be extinct, as will the horses, the birds and the bees, the animals and the trees, and the fish and the mammals of the seas. "And Oh! what a world this will be."

I cannot imagine myself as a child growing up without my father; his wisdom, his strength and his love. He was more important and needed during those years than life itself.

It is not the great depression my heart yearns for today; it is for t he way my father handled the depression years; and our lives through those difficult years. The 1930s.

And here I stand
All alone in an empty classroom
Where about me hang
Old maps, pictures, an abacus
Blackboards, and empty desks.

Into my heart an air that kills
From yon far country blows;
What are those blue remembered hills,
What spires, what farms are those?
That is the land of lost content.
I see it shining plain
The happy highways where I went
And cannot come again.

Remembering days those golden days of the past.

As I stand here in my old and forgotten
country schoolhouse.

Chapter 7

In the late autumn of 1934 my mother's father died in his home on the dairy farm in Hilliard. The hurt and the sorrow my mother felt for her loss of her father made her ill and very weak. My father took her with the baby, Henry, to the Vegreville Hospital where they both received professional care while he went on alone to comfort her mother, three brothers, and three of her sisters and to help with the arrangements and with the work that had to be done on their farm. He stayed with my mother's family on their beef and dairy farm for two weeks and on his return he brought my mother and their baby son home with him.

The sorrow my mother felt from not being at her father's side before he died, stayed within her and kept her weak for many years. A daughter draws happiness and strength from her father's love. Just from knowing that he is there. A father's love keeps her happy and strong and able to carry on with her own way of life, through the hard times and the good. When suddenly a father no longer is there, leaves a sad and lonely void, a weakness with no end. My mother's father, Metro Orenchuk was a kind and gentle man; orphaned as a small child in his homeland the Ukraine. As a very young boy he travelled all of Europe working at any job he could find, to support himself, his younger sister and a brother. In his early teens he travelled to Austria and met his wife, Maria Gach. Because he spoke fluent German; he then joined the German army. During a battle on horseback he was thrown and the horse fell on him injuring his upper back; which left him with a hump on his upper back for the rest of his life.

Maria's parents; Matthew Gach an Austrian, her mother Paranka who spoke both French and Polish fluently left home, Stojaniw in Austria, with eight of their children; seven daughters and one son. They travelled to Prussia in Germany where Metro Orenchuk joined them, and on January 11, 1899 they all travelled to Canada. Maria and Metro were married when she was 16 years old. Metro was in his early 20's. In Canada they settled on their homestead near Hilliard where Metro built a lovely three story mansion on a hill, with steeples, many big windows and a verandah. My mother Katharine was born there on December 15, 1908. The house still stands as solid and as proud as the man who built it in 1899. He was a poet and after he had built their first home in Canada he wrote this poem:

"Our House"

Наша хата мов палата
Хоч не псказненька
Деревана не богата
Все своя рідненька
і городець побічний
Ярина в ній гарна
Де садочок там веснуют
Мати господарна

My mother's parents raised three sons John, Andrew and Joe and four daughters Tillie, Katharine, Jenny and Agnes. Matthew Gach and Paranka were my mother's grandparents. They settled on a section of land also near Hilliard where my mother's grandfather built a European style house with a thatched roof and a mud floor. During the years, the eighteen hundreds, wars in Europe had people leaving their homelands, their homes and their families to the destruction of war, to settle in Canada, and to start building a new way of life. Great Britain declared war on Germany and her Austro-Hungarian ally in August 1914. Canada as a British Colony was automatically at war too. And on the 9th of April 1917, Tuesday midnight Vimy Ridge, a natural fortress held by the Germans since October 1914 had fallen into Canadian hands. This was the greatest victory that the British and the French had yet achieved in nearly three years of fighting.

World War I ended in 1918, and in 1922 the French ceded to Canada 250 acres of land on the crest of the Ridge for a memorial. Work on the 1st Canadian Division Memorial began in 1925, and was completed eleven years later at a cost of 1.5 million.The Vimy Memorial was dedicated on the 26th July 1936 by King Edward VIII.

Approximately half a million Canadians, out of the six hundred thousand who enlisted, served in France and Flanders during the First World War. Fifty-three thousand of them were killed in action or died of wounds or disease, and some thousands of others who survived were so badly mutilated, physically or emotionally, as to be dead in worldly terms.

Vimy Ridge is in the North of France, at the Belgium border. Flanders Fields, where the poppies blow, between the crosses row on row is in Northern Belgium bordering France and the Strait of Dover (the English Channel).

April 9th is a Veterans Day, a Commemoration date for the soldiers who died during World War I at Vimy Ridge in Northern France.

I was born on April 9th 1930, the Commemoration date of Vimy Ridge, 12 years after World War I ended and 12 years before the Massacre in Dieppe. Dieppe, south of Vimy Ridge on the banks of the English Channel in France, August 1942, World War II five thousand inexperienced Canadians were sent to combat to fight their kin by Britain's Winston Churchill and Major General J. H. Roberts. Three thousand of these very young Canadians were slaughtered before they could reach Dieppe banks. In history books, an Englishman by the name of Jack Harris, who's schooling had ended at the age of fourteen is celebrated as a Canadian who had done great things; the man with whom World War I began, and Vimy Ridge. Mother's sons, and daughter's fathers, were slaughtered; the healthiest and the strongest of men died in that war leaving the sick, the weak and the crippled to carry on. What great things did any war accomplish? But devastation; the killing and the rape of the innocent, the destruction of man's labours, lands, and nature, is what was achieved.

A Commemoration Date is to Remember

The house my mother Katharine was born in 1908. There was an outdoor oven built of stone, (a pyetz) outside the kitchen door. Mother Katharine, Anne and Eugenia Irene.

1993 - Mother Katharine on her father's homestead. The little shack out back, chicken house, cow barn and water pump.

Mother Katharine, Anne and Eugenia Irene in front of the Hilliard Post Office.

Mother Katharine and Joyce inside the house - 1993

Chapter 8

Many in the West, and especially during the 1930's, the ranchers and the farmers, could not afford a doctor, or long stays in a hospital. Peddlers regularly travelled the country roads with Watkins and Rawleigh patent medicines and combs, brushes, salves, remedies and cures of every kind for the ranchers, farmers, for their children and for their animals. Those peddlers were like travelling drug stores and more. My father's back was aching from his earlier years of clearing his land of rocks and roots, lifting logs for all his buildings and other jobs such as lifting the wagon boxes and the hayracks; placing them onto runners for the winter then onto wheels for the summer, lifting hogs and steers into the buyer's trucks. And when he worked at the paper mill in Washugal, Washington when he was 18 years old, he was named the "Paul Bunyan" because he was the strongest and most powerful. He even carried the logs on his shoulders. He never complained, and he never mentioned his aches or pains to his family and never went to see a doctor. And he never stopped dancing, though he walked with a twist. A peddler came by one time and he talked my father into trying Dodd's Kidney Pills for his aching back. Father took them through the winter months. They came in a little round box like a snuff-box.

When spring arrived and Father was out working the fields with his hired man. Mother was busy with her cooking and sewing and our nanny Annie Fundytus was in the pump house doing the laundry, my sister, two little brothers and I filled the pill boxes full of tiny green chokecherries and were swallowing them as we saw father swallow his Dodd's Kidney Pills. By late afternoon, Henry's skin and face was green and he was vomiting green, with a non-stop green diarrhea. Mother wrapped him into a blanket while father got the car to the door and I sat in the front seat with my father and when mother carried Henry out to the car, I held him in my arms as father drove at full speed, 50 miles per hour, all the way to the Vegreville hospital. After Henry was out of danger, he was left to stay in the hospital. Father and I returned home and found Edward having a convulsion. Father raced back to the hospital while mother held her first born son. Mother stayed at the hospital for a few days to comfort her sons. Anne and I stayed home with our nanny. I was a little bit sick but, Anne was fine.

When her sons were feeling better my mother came home and went to the "Icehole" (the lidonia) to get food. Anne followed and wandered off to the barnyard, although she was afraid of animals and seldom went out to the barnyard alone. A year old colt came galloping towards her to be petted and to be fed. When she saw him coming, she ran through the trees to where mother was, not using the pathway and stepped on the only rusty nail that was ever found. The nail went right through her foot. Mother had to pull hard and to pull to get that rusty nail out as it was attached to a small board and the board broke away. She boiled a kettle of water, sterilized a glass container and had Anne soaking her foot in sterile water with Epsom salts bought from the Watkins peddler, for hours. For the night mother then made a poultice of white bread heating some cream to boiling point and then soaked the bread in it. She placed this hot poultice around Anne's foot and wrapped it well. This home remedy was repeated and Anne had to stay in bed. This was quite a spring for my mother and father, with the two girls in bed being nursed back to health at home and missing school while the two boys romped around in the hospital. Dr. William Shandrow was the head doctor, he was an excellent doctor who loved children and all people from the very young to the very old. Anne's foot healed with no infection and the doctor did not have to see her. Our nanny, Annie Fundytus, was fired for not watching the children at play.

After Edward and Henry were brought home from the hospital; my father's youngest sister Paulina's daughter, Annie Hollowaychuk, was hired as our nanny housekeeper the summer of 1937. A happy jovial girl she was attentive, hardworking and she played a guitar. She wore a cowboy hat and she sang Country and Western songs. Anne wanted to play a guitar too and during our cousin's stay she taught my sister to play. Anne was then my accompanist when I played my violin.

Our garden was growing and the grain was green and tall in the fields. Leaving our hired man to milk the cows and care for the other animals, father packed us all into his 1932 Ford V8 model 18 car with a rumble seat with our nanny and away we went to Mundare where we picked up our grandmother and went to Elk Island Park for a camping week. We were all together, happy, and again all healthy. We had great fun playing in the sand, swimming, roasting wieners and marshmallows on wooden sticks over a camp fire. In a leaning birch tree we had our picture taken and we watched the buffalo roam. Elk Island Park was then, and is today, home for the few buffalo left on this land. A beautiful animal they are to see and to watch.

"Where The Buffalo Roam"
Oh give me a home, where the Buffalo roam
Where the Deer and the Antelope play
Where seldom is heard a discouraging word

And the skies are not cloudy all day.

These were the songs our nanny sang every evening as we sat around the campfire; "The Red River Valley Song".

From this Valley they say you are going
We will miss your bright eyes and sweet smile

"Beautiful Girl of the Prairie," was a favourite

Hi, hi, hi, hi,
Beautiful girl of the Prairie
With eyes of blue, and a heart so true
She's the girl I'm going to marry.

Hi, hi, hi, hi,
Beautiful girl of the Prairie
We'll ride along and sing this song
of our love that bloomed on the Prairie

"In An Old Dutch Garden"
In an old Dutch garden, by an old Dutch Mill
Where the moon was dreaming on a distant hill
When a smile danced by; It was then that I
Saw heaven in a pair of wooden shoes.

"I'm Thinking tonight"
I'm thinking tonight of my blue eyes
Who is sailing far over the sea
I'm thinking tonight of my darling
And I wander if he ever thinks of me.

"Happy Little Woodpecker Song"
He's up each morning bright and early
To waken all the neighbourhood
And bring to every boy and girl
His happy serenade on wood
Hear him pecking out a melody
Just peck, pecking on the same old tree
All day long.

"The Wreck of Old Number 9" was a song Annie sang and never will I forget her. Annie Holowaychuck with her older brother Nick and younger brother Eddie, were orphaned when they were small as her mother Pauina died giving birth. They had to cross the Saskatchewan River to get to a doctor and did not make it in time. Her father died soon after from the sorrow of losing his wife and the baby. I believe he blamed himself for not getting Paulina to a doctor in time.

Annie's father, John Hollowaychuk, was an engineer, on trains along the railroad our grandfather Michael Yawney was building until he died in 1905, the year my father was born. When Annie sang this song, with tears in her eyes; we all cried.

"The Wreck of Number 9"

On a cold winters night, not a star was in sight
and the North wind came howling down the line;
with his sweetheart so dear, stood a brave engineer
and his orders to run old number nine.

He whispered goodbye with a tear in his eye,
and the joy in his heart he could not hide
Oh the whole world seemed bright, for she told him that
night, that tomorrow she'll be his blushing bride.
The wheels hummed a song as the train rolled along
And the black smoke came pouring from the stack
and the headlight a gleam seemed to fit in his dream
For tomorrow when he'll be coming back.

He sped round a hill, and his brave heart stood still
for a headlight was gleaming in his face;
And he whispered a prayer, as he threw on the air
For he knew this would be his final race.

By the wreck he was found, lying there on the ground
And he asked them to raise his weary head
As his breath slowly went and this message he sent
To the maiden who thought she would be wed.

There's a little white home, that I meant for our own
Where I dreamed we'd be happy by and by.
So I leave it to you, for I know you'll be true
Till we meet at the golden gates - Good-by.

Chapter 9

1937 was a very good year. The grain grew high and farmers were receiving more money for their grain and their livestock. In the spring of 1938 father bought a new gunmetal grey Plymouth car. We were all so proud of it, as it was modern and we loved it. There was a feeling spreading over the land, that the hard times; the Great Depression years were slowly drifting away.

My father had been planning to take us all to the Rocky Mountains and in June 1938 when school was out, father packed us into our new Plymouth car and away we drove our familiar old country road to Ranfurly, then Vegreville and on to Mundare to pick up our grandmother then off to Banff in the Rocky Mountains. This journey was exciting and we had not seen mountains before this. No one we knew in the Lornedale School district had ever travelled so far. In Banff my father rented a little house, for a month. It had an upstairs and a veranda, it was painted creamy yellow with green trim around the windows and the doors. It was fully furnished including bedding, linens and dishes. There was a free standing bath tub and a little hand pump in the kitchen sink. There was an old country wood stove with a hot water reservoir on the right side. In the yard were tall evergreens, a small vegetable garden and flowers, and the picket fence was a creamy yellow matching the colour of the house. There were squirrels, chipmunks and birds of every kind. Wild deer, and the elk came right up to the picket fence and we could touch them.

After a light breakfast we were off to the sulphur hot pools. Father looked after his two little sons while in the pool, and mother kept an eye on my sister Anne and me. Anne and I were hanging on to each other and bop, bop, bopping along when we bopped once too often and the water was above our heads and we both swallowed sulphur water. Mother pulled us back out by our hair which was floating on the water.

My grandmother did not go to the sulphur pool often, she did all our cooking and she'd sit out on the veranda and sew. She hand sewed cushions, tea towels, towels, pillow cases, and her long peasant skirts, long sleeve blouses and shawls. Father insisted she come to the hot sulphur pool with us a few times. He believed hot mineral water had healing powers, and was relaxing. Our rented bathing suits were like long john's; full length from ankles to neck with long sleeves, and were all brown.

Grandmother always had a big meal ready for us when we returned; fried chicken, vegetables, cabbage rolls and more. But when grandmother came to the pool we all went to a restaurant for lunch or we bought hotdogs.

We went for long walks, along trails through tall trees with bridges over streams, and we watched waterfalls. We watched deer, elk, bear, and saw moose and buffalo. We climbed trees and mountains. We watched an Indian Pow-wow. The Indians came walking and on horseback, dressed in their native clothes of leather, beads and feathers with moccasins. Young mothers had their babies strapped to their backs in leather cradles. Some elders were pulled along behind a horse with a buffalo hide stretched between the two poles to lie on and carry their wares. They beat their drums, danced and sang all that day. And when night came they all quietly walked away. Disappeared into the night.

We were afraid of the Indians as we had not seen any before. The Royal Canadian Mounted Police, so handsome dressed in their red coats and Stetsons, looked so proud and firm, but we were afraid of them too.

We saw gardens of beautiful flowers, shopped in the many stores and bought souvenirs and we went to a museum.

The month went quickly and we packed for our long drive home, back to the farm.

My mother sulked the whole time we were in Banff as she was not happy that father brought his mother with us. Mother's indignant feelings towards my grandmother lasted until her death and beyond. Even 20 years after my grandmother died, whenever she was mentioned my mother made unkind remarks. I can still see my grandmother; a dear, gentle, quiet and caring, strong physically and mentally, and most of all; an independent great lady she always was, and I will forever remember her and miss her.

Father wanted us to enjoy a holiday but he was also searching for a way to ease his aching bones. And he was only thirty-three.

We did not know he was in pain. Through all his suffering with aches and pains he was always happy and caring. He loved his sons and daughters, and Katharine was his whole life. She was his world. Never a harsh word was ever spoken to her. He cuddled and comforted her through all of her emotions, his whole lifetime.

I feel sad because we, his family did not give him support and strength when he too needed some. We did not know.

Through all our lives with father he was our pillar and our strength.

Though leaves are many the root is one;
Through all the trying days of my youth
I swayed my leaves and flowers in the sun;

Nanny Annie Hollowaychuck with Anne, Irene, Edward, Henry and cowboy hat on ground

Elk Island Park - leaning birch tree Irene, Anne, Edward, Henry, Dad, Mother, and Grandmother

Eugenia Irene with violin, Anne with guitar

Elk Island Park buffalo

~~1962~~ 1938 - Father in Banff

1938 - Banff Holiday

Now I may wither into the truth.

MUNDARE

Mundare was my grandmother's home, she loved that town, the people there and her cottage next to the little church my father and mother were married in. As a little girl of three to four years old I stayed with her during the summer months. There was a stile to get over a fence from her yard into the churchyard, and every Sunday we climbed that stile to the church services. Grandmother held my hand as I climbed up and over, then she'd come over holding her long Sunday's best peasant skirt wearing a pretty blouse and shawl, a fancy hat, gloves and hightop boots, looking very prim and proper. She'd then take me by my hand and off we'd go to the church service.

In a large duck pond across the road from grandmother's cottage were many ducks and frogs and every day she held my hand as I skipped along beside her across that road to the duck pond to feed the ducks with bread crumbs and garden greens she had stored for them and there we watched the bull frogs leap, swim, then bask in the warm sun, catch flies and croak. After feeding the ducks, I'd skip along beside her into town to get candy while grandmother had her visit with the townspeople awhile.

My grandmother always grew a large garden and she raised many geese, ducks and chickens. She taught me how to dress and prepare the birds for the market and for the oven. One day she left me alone in the cottage preparing the birds while she went over the fence, along the churchyard to a well which was still there from the homestead years, to bring water for us in the cottage. She had a pole that fit over her shoulders; this pole was called a shoulder yoke, it held a bucket at each end. When she returned with the buckets full of water, she found me hiding behind the stove wiping tears off my face, and she asked "Were you crying?" I said "Oh no! I was putting some wood into the stove and smoke got in my eyes." Though really I was crying. I was homesick and did not want her to know that. She did know, she stopped all our work and she held me close to her while she told me stories of her life as a pioneer, and of the life my father had while he was growing up on her old homestead north of Innesfree without her and without his father.

When I stayed with her, everything she did in her home and garden I wanted to do too. I wanted to help her in every way that I could. One day without telling her, I went out to her koomora (the shed) to bring back some coal for her stove. In that koomora was a broken bucket with no handles. I filled that bucket with coal not realizing the weight of it and when I went to pick it up with both my hands, it was much too heavy and I fell on top of it cutting my lip through to my teeth. I left it all there and returned to the cottage where grandmother immediately wrapped my mouth with bandages around my head and neck. I could not talk or eat. She fed me liquids through a real straw, a wheat or oat straw and she took me into town every day so that I do not even try to talk but to listen and I was to do no more work. With my mouth wrapped tight, listening and being interested in what people there were talking about got my mind off of what had happened and my cut lip healed in no time. When my father and mother came to take me home, I was able to tell them everything that had happened and of how grandmother healed my lip.

I was five to six years old when I had my last visit with my grandmother. I was in school then and was able to be of more help around our farm. During this last visit, I asked her to tell me more stories about her life: She told me that after Mike died she became very ill, and was not able to get out of bed, that all her strength was gone, and she was not able to light her lamp, or get any food. Though there was a small tin of sardines on the window sill beside her bed, (she had a homemade feather bed in her kitchen) she was able to open that tin of sardines and she nibbled on a tiny piece of a sardine every day, that was all she had. This tin of sardines she said lasted her the duration she was ill. This window beside her bed faced towards the churchyard and there were many trees between her cottage and the church. The church people did not know that she was there and did not come. The townspeople would look for a light in her window and when they saw no light; the townspeople believed that she went away to stay with Julianna and Paul. Julianna and Paul had built her a new home on her first old pioneer homestead for her to go back to, that they would be there to care for her. Being the independent lady she always was she was not going to burden anyone with her aging years and she would not leave her two room cottage in Mundare. No one came, she was all alone, there were no telephones in homes during the early Canadian developing years. There were no strings to pull as we have today if we can't get out of a bathroom, and no buttons to push to call for help - A Life Line - A Help Needed Alarm. No one came. After a month passed she was getting her strength back and was able to get out of bed. She did not know what had happened and why she was so very weak.

Remembering my grandmother and knowing her to be a strong and determined woman, I believe her determination and willpower pulled her through when she regained her strength, and today I believe that she may have had pneumonia and somehow, on her own, nursed herself through it. She believed the oil and sardines, when they became rancid, were medicinal and helped her to survive.

My grandmother had a coffin in her back room which she had purchased for herself, it was neatly covered and had family pictures and artifacts placed on it. It was completely disguised and she also had her own funeral service all paid for.

Remembering my grandmother, my father's mother, never ever had I ever known or have I ever heard her talk against someone or complain about anything or anyone. Or of the difficult life she had in Canada, which I am sure that it was - most difficult. Alone, giving birth to her first son in 1905, and in her home nursing Michael the baby's father dying of pneumonia after all his money was taken away. Then Mike: raising his five sons and nursing Mike through most of their married life till his death. To me she always was and will always be remembered as a kind, gentle, real lady with a true devotion as a wife, mother, mother-in-law and grandmother. Devoted to her country Canada and to her marriage vows till her death. She died August 18, 1961 and was put to rest in Buchach which is the name of the cemetery near her first homestead home where Michael, my father's father, is buried. The newspaper said my grandmother was 88 years old when she died, though that is not known for sure.

Grandmother's Koomora 1952

Mother Katharine, Anne, Irene visit Koomora 1993

1993 - Anne beside the house she was born in, June 14 1928

1955 - Grandmother's cottage - window looking toward town of Mundare. Church in background

1993 - New church replaced the old

1993 - Museum where my grandmother's cottage was. The koomora sits in front of the museum today

Chapter 10

Our holiday in Banff was interesting, exciting, and most of all relaxing. We learned much about a different way of life, about different animals and of people with different lifestyles; different from our familiar surroundings and sheltered way of life on the farm; it was as if we were in another world for a whole month.

When winter arrived that year 1938-1939, the strong winds blew the deep snow into drifts covering fences, farm machines and sheds. Horses and a sleigh were used for all traveling; into town for supplies, and to the school. Dance parties were held at the school whatever the weather and always I went along to play my violin. Earl Hardy our school teacher played the piano and anyone else who could play whatever else always joined in. The school was also a church, a concert hall, a bingo palace, a hall for card parties and whist drives, grain growers meetings and school board meetings.

At each dance party a pie social or a basket social was held. At a pie social, each lady arriving, married or single, would bring her favourite pie to be auctioned. At a basket social each lady brought a complete lunch packed in a decorated basket to be auctioned. The auctions were always held at midnight and the lady who made a pie or packed a lunch basket was to share it with the man who was the highest bidder. After they lunched she had the first dance with him. Then they could continue dancing the night away, or each go their separate ways. The Home Waltz was always played at three in the morning.

It's 3 o'clock in the morning
We've dance the whole night through.

The money collected at these socials was for the school supplies, books and desks.

Spring in 1939 came early with Chinook winds softly blowing in off the Rocky Mountains melting the snow. Then, suddenly, March came in like a lion with a cold wind, and a frost made all the slush solid ice and our yard and the roads became slippery ice. The cold formed an ice crust on the snow drifts still piled high over the fences and in the fields, we walked over the fences on top of the snow to get to school. Our horses and cows had to be kept in the yard where they were fed hay and chop, and were watched and protected from slipping and breaking a leg. The ice crust in the fields would cut the horse's and the cow's fetlock, (their ankles), if they walked through ice covered snow. Even my father put nails into gunny sacks, then he tied the sacks to his boots to walk out into the yard.

A baby was due to arrive in March, but not knowing the exact date and because of the icy roads, father hitched his strong and heavy matching black team of Clydesdales, Duke and Beauty to a sleigh. He picked mother up at the front door where she was waiting; and away they went, the sleigh slipping and sliding from side to side along the driveway and on to the country road, Duke and Beauty proudly clippity clopping along. They were off to Vegreville to await the birth.

On St. Patrick's Day Patrick was born. When father stopped his team at our door we all ran out to see the new baby and on a bed of soft sweet smelling hay, all wrapped and covered in down quilts and pillows; mother uncovered for us the beautiful baby. He had blonde curly hair and dark blue eyes. He looked at us all with his big blue eyes and smiled. The nurses had named him Patrick, so we added Eugene and Michael, which gave him three names. Later that summer mother and father went to Mundare and Hilliard to visit relations and to show their son Patrick. After they left, Edward and I went to the pasture to pick wild strawberries but first we stopped at the school to watch the spring water run into the meadow and where we took a drink.

I'm going out to clean the pasture spring;
I'll only stop to rake the leaves away
(and wait to watch the water clear; I may):
I shant be gone long - you come too

Then to the end of our pasture we ran to pick wild strawberries, tiny, very sweet and full of flavor. This strawberry patch was on a little hillside, on the west side of thick trees bordering our country road. As we were busy picking, I noticed one cow alone, amongst the trees while the rest of the herd were at the north end of the meadow. I left my little bucket of berries with Edward, and while he continued picking and eating berries I ran off to see why the cow was alone and there beside her was a newborn calf. As I stepped closer to take a look, she turned and came after me. I ran as fast as I could across the stream to that tree in the meadow. When I stopped the cow stopped too and then turned and ran back to her calf.

And forever I will remember
There's a tree in that meadow with a stream drifting by
Where I was safe from the cow protecting her calf

After the cow was back with her newborn calf and hidden in the trees, I ran along the stream to the school yard where I met Edward with our picked berries, and from there, we both ran home.When father and mother returned we told father about the newborn calf, though we did not say that I was chased by the cow.

I'm going out to fetch the little calf
That's standing by the mother. It's so young
it totters when she licks it with her tongue.
I shant be gone long - you come too.

Father hitched Percheron Pete, our dapple grey horse, to the stoneboat to fetch the newborn calf and when he returned the cow was following behind, excited, and mooing all the way home. They were kept in the barn to be safe from any wolves who might be sneaking about. The cow's name was Cranberry, her colour was a dark brown-red of the Red-Pole breed and why we named her Cranberry. In the morning father let Cranberry into the yard with her calf. He was in the pumphouse working the cream separator, when Cranberry suddenly headed for the pasture, her calf stumbling behind. I saw her leaving and quickly ran out to bring her back, Cranberry remembered me from the day before and she turned and came after me at full speed and as I was running away, down the little hill towards the pumphouse she hit me and sent me flying. I rolled as she jumped over me lengthwise and all I saw was a big fat belly, four teats flopping and four big legs flying over me. She stopped with her hind hoofs on each side of my face. Then she quickly turned, and with her big nostrils touching my face she snorted as she looked at me with her big brown eyes then she ran back to her calf. All I could think of at that time was;

Hi diddle diddle
The cat and the fiddle,
The cow jumped over the moon
The little dog laughed
To see such a sport
And the dish ran away
with the spoon.

Before I was able to sit up and realize what had happened, father was there beside me. He saw from the pumphouse window what had happened and Cranberry was put into a corral with her calf until the next cattle auction where they were sold. Father had bought Cranberry at an auction in the autumn, after we returned from our holiday in Banff. He was not aware of her moods with children, and she was fine before her calf was born. All the other cows were tame, they accepted us milking them, riding them, and playing with their calves.

My mother's grandmother Paranka Gach died in June 1939. Her grandfather Matthew Gach had died in the 1920's. His homeland Austria, was taken over by Germany in 1938, and on the 1st September 1939, Germany invaded Poland. Before both my mothers grandparents died; they could not have been aware that their relations would be at war with each other. My mother's grandmother was a Polish French, her grandfather an Austrian.

September 1st, 1939 - World War II had begun. Mixed emotions were felt by all; there were smiles on peoples faces, with an extra energy in their walk and talk - the depression

years were at an end. Yet a sadness appeared in their eyes; a sadness of not knowing what lies ahead. This was the beginning; the beginning of a world war. Not a leaf stirred . Not a breeze was felt. Not a song of a bird was heard. I believed that for a moment the whole world stood still.
Thunder, lightening and hail the size of golf balls came pelting down on our ripe grain, pounding it all into the ground. Father, mother and our nanny held pillows against our windows, to keep breaking glass from crashing inside. After the storm blew over, not a stalk of grain was left standing. All our grain was pounded flat, and covered white with ice balls. Everywhere, everything looked like winter in the summer. Very few farmers had grain to harvest that year. My father sold more of his hogs and shorthorn steers to keep us fed for the winter and there was not much garden to bring in to store for that winter. But the loss did not keep the growers from having fun. A pie social was held, and the Lornedale school was a-rocking and a-rolling with neighbours dancing the night away to the music of my violin and Earl Hardy, my teacher, on the piano. My mother had a pretty silk dress made for her by her tailor cousin Bill Zaruby in Vegreville and she would wear that dress to the pie social. She made father's favourite pie, a lemon meringue, and a pie for me to be auctioned. She knew father would buy her pie whatever the price. And father did. He was happy she made his favourite pie.

I sat in the back of our Plymouth with my violin beside me, holding my pie on my lap. Mother was sitting in the front seat beside father looking very pretty in her new silk dress; holding her lemon pie piled high with golden meringue, on her lap. And away we went driving along the country road towards the school, and we were singing:

"Pack up your troubles
in your old kit bag
And smile, smile, smile."

Mother seldom showed signs of being content, and she seldom smiled. Father was always playful and when he wanted her happy and relaxed, and in a party mood he would give her a "happy-love-slap" to her lap with a gentle "squeeze to-her-knee" then she would giggle. And Whack-O-Obbliga-to, the lemon and the meringue were all over their faces. In their hair and all over Mother's new silk dress. The horrified, surprised look on Father's face covered with lemon and meringue, and the tears Mother shed; (I'm not sure if the tears were for the mess on her dress, or for the loss of a lemon pie?) Father pulled the car over to the side of the road to comfort Mother with his hugs and his kisses, spreading more lemon and more meringue on their faces. Oh ! What a sight to behold. This was early winter 1939; Father was watching the road as we were driving along; not the pie. We drove on past the school yard to the crossroads where we turned around and drove back home. (Someone should have been waiting there with a camera; this would have made a "Believe-it-or-not" Pie Social poster of the year) After Father washed and cleaned the car, Mother changed her dress and took with her another pie; we were off again, to the dance party and Pie Social. Paul Kryzanowski was playing his violin for the dancing when we arrived, a little bit late. The dance tunes at these dance parties were "When I Grow Too Old to Dream", "Blue Eyes Crying in the Rain", "Beer Barrel Polka", "Turkey in the Straw", "My Little Redwing". "Pop Goes the Weasel" was played to end a dance, between dances.

All around the carpenter's bench
The monkey chased a weasel
That's the way the money goes
Pop goes the weasel.

After all that had happened that evening, we all had great fun. It was, and always will be, a night I will forever remember.

Chapter 11

The Second World War continued and hired help was impossible to get as young men and woman found work in the cities or joined the armed forces.

Our nanny Annie Fundytus was no longer with us as Anne my sister told my mother she noticed her showing a special interest in father. Believing her, mother sent her away. Most of the nanny's we had were very young; and were often homesick. They looked to father for acceptance and approval; a father figure. Mother did not tolerate this. They were to stay away from father. Through all the years I knew my father he was an honorable man, with the highest of morals. A man of virtue and true grit. He was fun loving, compassionate, he cared for all living creatures. And he trusted people. My mother was not like that. She was selfish, possessive of her spouse and always complained she had too many children and too much work. As we were growing up she convinced my two sisters, but not me, to shun father. I know this must have hurt him very deeply. Always he respected mother's feelings above all else and soothed her moods so as not to hurt or upset her or make her cry. He seemed to understand she was being tortured by her own false feelings.

I believe jealousy and possessiveness are hereditary and incurable. If instilled into the very young it can destroy their lives from generation to generation. It can cause confusion, hatred, divorce, mistrust without rhyme or reason. I too was bit by that jealousy bug, therefore I do know the agony of it. My sister Anne grew up not really knowing a father's love.

She stayed away from him as my other sister did too. They followed mother's wishes.

Edward, in his early 20's, met a young girl, his first love, who was very much like he was. They both liked horses and country music. Her name was Darlene Kemp. Her brother Evan Kemp was a wellknown country and western singer and songwriter. Darlene also sang country songs, and played a guitar. Edward played many instruments and he often went along with Darlene and Evan Kemp to entertain at private parties and barn dances. Darlene grew up without a father; Why? I do not know. Darlene Kemp loved my brother, and she adored our dad as a father figure. Mother did not tolerate Darlene's admiration towards father and she put a stop to Edward's romance and Darlene was no longer welcome in our home. We never saw her again.

Edward married a young teacher, Diane Carter, an only daughter who had two brothers. She was a young, immature girl who was loved by her parents. Edward, a first son, was cared for, treasured and given everything, plus more, by his parents. This marriage was doomed from the beginning. Edward and Diane were too much alike to build a solid marriage. They had two sons and a daughter. Their three children were very young when that marriage ended. Diane found another love on a Greyhound Bus.

Henry, a university student, at 21, married a happy dancing red-headed telephone operator named Joan Mathieson from a large family who showed no interest in father. All her love was directed at Henry. They had eight children - five boys and three girls. Joan became mother's friend for life and Joan never complained about having a hard life or too much work or of having too many children. Henry and Joan's marriage did not last either. Henry was the jealous one.

Oh jealous heart
Oh Jealous heart stop beating
Can't you see the damage you have done,
You have driven him away forever
Now I'm the lonely one.

All through his growing and developing years, Henry was kept in the background. He was not held close, nor was he cuddled as a growing child should be. He was blamed for being the baby of over 12 lbs. at birth who gave mother a bad time without a doctor present. Because Mrs. Gilpin, R.N. was there from the early beginning, and the doctor from Viking did come later, was never mentioned. Many years later, after his divorce, and as a middle-aged man, Henry showed resentment towards father. Could this have been his sub-conscious mind telling him of his lost forever desire to have been held, loved, and accepted; which never was. His child emotions broken, Henry matured into a big handsome man of over 200 lbs. with a genius brain. All through grade school and University he was an "A" student and he never did homework. Always, he was a most hospitable gratuitous host. He loves his family, and he loves people, the intellectuals.

Anne, the oldest, married a "tall-dark-and-handsome, resourceful" Australian Paul Chauvel, a fisherman who loved her with a never ending love. Together they raised two sons, but her jealousy and mistrust drove him to drink and then she divorced him, and lived common-law with a lazy, good-for-nothing Don Juan for a time.

Now me - Eugenia Irene. Well! I am a complete story, which will be told in my next book. I was married twice and have one precious son; and I have had a walking disability since I was 30 years old. This is due to a doctor's identity mistake.

If you keep your eyes on the rainbow
You won't mind the rain

To live with an accidental disability from a young age is an agony beyond words.

One must live it to understand it.

Jealousy festers into anger, hate, mistrust, and is often misunderstood. It takes a powerful and understanding individual, with a strong body and of mind, to live with, accept, and be in control of and tolerate jealousy for a lifetime. To shun the heartache of jealousy, is to just walk away; as many do. They leave their spouse, their sons and their daughters without love nor money. Some, because of jealousy, drink themselves into oblivion and get lost to the world. The more dangerous ones, are the violent ones, they get out of control with their anger, and they destroy - because they do not understand, nor do they want to understand that insane emotion, jealousy.

My father accepted his Katharine for what she was, and for the way she was. He did not try to change her. He loved her completely, tenderly and emotionally with his whole heart and soul. My mother was loved like no other woman has ever been loved by a man and all through her life with father, she was not aware of that. Because of his devotion to his Katharine. Did my father stop growing as an individual??

I believe that he did.

However, I am writing this book about the greatest man that has ever lived. He was not a Mayor, a Premier, nor a President. He was a real man.

A daughters pride.

Chapter 12

In harvest season, on September 8, 1940, Marshall James was born with white hair, hazel-brown eyes and a dimple in his little chin.

Annie Hollowaychuk, who once had been our nanny, came to help with the threshing with her brothers Nick and Eddie. Nick brought the girl he was going to marry; for father's approval and for us to meet. She was a beautiful blonde blue-eyed girl and Nick was six feet with black hair, a black mustache, a twinkle in his eye and a song in his heart. An image of Michael Yawney, my father's father. They were baking pies, frying chicken, cooking breakfast and serving our threshing gang. Eddie was strong and fair haired with broad shoulders and was in the fields every day pitching bundles into the threshing machine.

Annie, Well! It was like having her back home again. We were all very happy she came. She knew what had to be done and she did everything to help and after supper she entertained us with singing and guitar playing. My sister Anne and I joined her and we all had a happy relaxing sing along. When the threshing was all done Annie, Eddie, Nick and his bride-to-be left to plan the wedding. Family weddings in Alberta could be a two week celebration.

Nick and his bride were married in Mundare in the same little church my father and mother were married in, in 1926 and their two week reception was held on my father's old homestead, his birthplace. My sister Anne, aged 12, was one of their bridesmaids, she looked so pretty in a long pink gown with pink ribbons in her dark hair. She was excited and happy, and so were we. We all stayed at the reception on father's homestead for the two whole weeks and had fun.

Soon after the wedding, Nick went to war leaving his bride alone and lonely and she stayed with my father's family for a time, before she found work in Edmonton. Several years passed before Nick returned from the war but his bride had died. The anguish she bore was more than her gentle and troubled heart could take and Nick just faded away and we never again saw him.

1940 was a difficult year for my mother with two little babies in our home and with no outside help. Patrick was one and a half years old when Marshall James was born. Henry was just starting his first year at school, and Anne, Edward and I were still all at school; and nowhere was there any help to be found. Not for the farm nor for the home. Though, I expect, the farm was a better and healthier way of life than for the families who lived in the cities. In the cities people were put on food rations; they were all given coupons for the different foods they could buy, and the portions were all very small. This food ration lasted until after the war ended.

It was a bad year in 1941 because of the drought, the crop was poor. As father was cutting the grain, my sister Anne and I were stooking and we were almost keeping up to father's binder. Every time he finished a section he would help us before he went on to another field. That year we did it all.

Mother came every afternoon at tea time, with the baby in one arm while dragging the other and Edward and Henry carried our mid-afternoon lunch. We would eat in a nice cool place under some trees, but sometimes we would spread our picnic lunch beside a stook if the trees were too far away.

1942 the 4th of July, Joyce Marie was born in the Viking

General Hospital and mother nearly lost her. Joyce Marie was tiny with black hair, dark, dark eyes and dark skin. Mother told us that she could not believe that she had a baby with black hair when her previous babies were all blonde or bald. For a time in the Viking hospital she said she thought some babies had been switched but after she inspected all the other babies she was then convinced that Joyce Marie was ours.

That summer with no hired help we had to do all the work ourselves. Anne and I again did all the stooking. We milked the cows and we separated the cream and churned the cream into butter and made cheese. Paul Kryzanowski, who owned the threshing machine, with my father's help and that of one other neighbour, were doing all the threshing, just the three of them. They threshed one quarter on each of the individual properties they owned but before the second quarter could be threshed, it snowed. The grain that was not threshed was covered in deep snow until spring

During these years because of a shortage of outside help Anne and I did the janitorial work at the school. We swept the floors every day, emptied waste paper baskets and dusted the blackboards and the desks. During the winter after we milked the cows in the early morning we were at the school to light up the big round wood and coal furnace in the basement so the school was warm for the teacher and students, and then we returned home for breakfast often with our noses and toes nipped by frost. We would hold a snowball over our noses and put our feet into snow to thaw out. Then, I had to return to the school. Anne had completed grade 8 in June 1942, before Joyce Marie was born, so she became our fulltime nanny. She did all the housework and looked after the babies. I helped father out in the yard feeding horses, cows, pigs and chickens and I helped mother with all of her cooking and baking. I loved to cook and to bake, and I did it well, making bread, pies and cakes.

From 1926 and through the depression years to 1943, and their life on the farm, seven children were born:

Born June 14, 1928, Anne, the oldest daughter, was a studious type. She liked books, reading and studying. She said she wanted to be a teacher when she grows up. She was afraid of farm animals and of being home without elders. She grew up to be 5' 7" tall, beautiful and slim with dark hair, blue-grey eyes and dark skin. She could have been a model, but instead she became a librarian. She married an Australian and raised two sons.

Born April 9, 1930, Eugenia Irene 5' 4", fair haired with hazel eyes and white skin, at three years played the violin and at age five I played my first violin solo at a Christmas concert. Later, at age sixteen, I studied in Vancouver and played with the Vancouver Philharmonic Symphony Orchestra. After winning a cup for the best violinist in the Okanagan, I travelled to study at the Royal Conservatory of Music in Toronto from 1949 through 1955. I married in Toronto and raised one son named William, a first born grandson.

Edward, the first-born son, was born March 26, 1932, slim, tall, fair haired with blue eyes and white skin. He loved the farm and animals. Mother's and father's pride and joy became a school-teacher, then a school principal. Edward always had a ranch of Clydesdales, thoroughbreds and quarter horses, all for show and riding. He taught music and plays almost every instrument that was ever made - violin, piano, banjo, guitar, harp, mandolin and many more. He married a schoolteacher and raised a daughter and two sons.

Born April 28, 1934 with dark hair, dark, dark, eyes and a dark complexion, Henry was a giant of a man with a genius brain. He never studied, never had to. An A student all through school and university, Chemistry and Pharmacy were his majors. Henry married a local redheaded telephone operator. Together they raised eight children, five boys and three girls.

Patrick was born March 17, 1939, handsome, tall, slim, black haired with blue, blue eyes and white skin. He became the Okanagan Valley property tax assessor. He has a pleasant personality and a quiet disposition. Patrick married a local girl and raised three sons.

Born September 8, 1940, Marshall had light brown to blond hair, hazel eyes, dark skin, a dimple in his chin and was over six feet tall. He had a PhD degree in sciences and languages and made his home in Victoria. He married a local teacher's daughter whose father is an opera singer. Together they raised one son and a daughter. The whole family plays musical instruments

On July 4, 1942 Joyce was born with dark hair, dark eyes and dark skin. She is a small 5'2". She had a university education that led her to work in the field of travel agencies and airlines. Joyce married a handsome, compassionate, kindly man from Calgary (not a cowboy). Together they raised two daughters.

We were a happy family and a farm was the best place to be during the depression years.

Father talked of university educations for his sons his daughters, and he knew that mother was never really happy with her life on the farm.

1943 - The beginning of the end.

One morning, March of 1943 our stooks were still deep in snow and my father deep in thought packed a suitcase, gathered us all for a hug and then with suitcase in hand slowly walked towards the barn. He looked at the trees, the fields and the buildings. He talked to Hero and to Rex, his two faithful dogs. He stopped and he talked to all his farm animals; he then looked at his whole world around him. In the house, we stood there watching him with our noses pressed against the window pane. My mother was crying, the baby was crying and I was crying too because he was leaving. He went into the barn, harnessed Duke and Beauty, his strong Clydesdales, he hitched them to a sleigh, and alone went down the country road to Ranfurly where he left his horses in a stable to be cared for and there he boarded a train going west.

The train took my father through the mountains and to an Apple Valley.
What made my father go there?
What was drawing him into mountains and to an Apple Valley;
to a place where he had never been?
Was this because his father; a father whom he never knew, came from an Apple Valley and the far away majestic mountains of Asia;
was this his fate?

У горах Карпатах
Ой, Там би я жив
Згори на долину
Все би м ся дивив

Там пташки співають злегенька
Все, уха-ха-ха
А голос сопілки там чути щодня
Там пташки співають злегенька
Все, уха-ха-ха
А голос сопілки там чути щодня

У горах Карпатах
Там сосна росте
Там пахуче зілля
Гарненько цвіте

Там пташки співають злегенька
Все , уха-ха-ха
А голос сопілки там чути щодня
Там пташки співають злегенька
Все, уха-ха-ха
А голос сопілки там чути щодня

What follows is a rough translation of the Mountain Song he remembers of his father:

Oh! Mountains majestic Mountains
Oh! That is where I long to be
From Mountain tops to Valleys below
I will forever admire and see
Where the birds sing so sweetly
and laughter echoes everywhere
Where the song of a Meadowlark
is heard every day

Oh! Mountains majestic Mountains
Where tall evergreens grow
Where the air smells of perfume wherever you go
From blossom of wild flowers and orchards below
Where the birds sing so sweetly
and laughter echoes everywhere
Where the song of a Meadowlark
is heard every day

Chapter 13

At the end of March father returned after three weeks and as we gathered around him he said, "Get ready to move! I have found the Garden of Eden."

For fourteen years this farm was his world. Six children were born here. The buildings he built, the land he developed, with this history around him, would it be easy to leave?

We quietly sat around him and listened as he told us of his train ride through the Rocky Mountains. He said that early one morning a green valley appeared. This was the Stepney Valley. He saw herds of horses, cattle and sheep. White wooden fences separated the farms. The farm homes among trees had white picket fences. So beautiful, he said. The train was moving along slowly, puffing black smoke. As father was admiring the serenity and beauty of this green valley, the train stopped. This was Armstrong. father checked himself into the Armstrong Hotel and contacted a realtor and soon was exploring the surrounding area.

This little village was incorporated into the city of Armstrong with a population of 1,200, on a Saturday, 19th of April 1913.

The place where Armstrong now stands was the hunting grounds of the Shuswap Indians. This tribe in various bands occupied the land between Larkin, Sicamous, Salmon Arm, and Mable Lake. The Shuswap Indians were probably the most fortunate in the whole of Canada; at no time did they suffer what is known as hard times. The lakes and the rivers always had their Salmon and Kokanee runs. Deer roamed the forests in herds and the valley was the biggest chokecherry and saskatoon patch in British Columbia. Never was there a shortage of dry meat, dry fish and pemmican. All this with a very pleasant climate. The year 1811 the white man came with their superior intellect and firearms and took over their fertile valley.

In 1892, the Armstrong Station received its name in honour of "William Charles Heaton-Armstrong", Lord of the manor of Roscrea, Ireland, a London Capitalist who had bought bonds of the Shuswap and Okanagan Railway; a spur line connected Sicamous with the Okanagan Landing; four miles south of Vernon. The S & O Railway began operation in June 1892.

In addition to the hospitable Armstrong Hotel, was a large cheddar cheese factory the famous Armstrong Cheese collecting milk from local farmers. There were schools, a newspaper, The Armstrong Adviser, The Armstrong Light and Power Company, the Armstrong Spallumcheen Hospital, the Armstrong Co-operative Store, a garage, a sawmill, the Bank of Montreal, Buckerfields and more. Father was excited and happy as he sat in the Armstrong Hotel with the realtor writing an agreement for him to purchase acres of grassy meadow with land for growing celery and a huge three story white house on a hill overlooking Spallumcheen, Stepney, and all of Armstrong, when a train pulled into the Armstrong Station going South. Suddenly the train became a beacon; it was calling him; drawing him; telling him to go on. Before he signed that real estate document to purchase; he checked himself out of the hotel; he hopped aboard that train; and twelve miles south of Armstrong; there it was; the beautiful, sunny Okanagan Valley. A fruit growing valley in the mountains.

This, he found was to be his eternal home; his

"Apple Valley"

AUCTION SALE

HAVING RECEIVED INSTRUCTIONS, I WILL SELL BY PUBLIC AUCTION FOR

WM. YAWNEY

Who is leaving the district and going to B.C.

at his farm, the N.E. Sec. 29-49-12

3 miles west and 11 miles north of Viking: 1 mile west, 10 miles south of Ranfurly

Tuesday, April 27 1943

Sale Starts at 12.30 o'clock

Free Lunch at Noon

the following Goods and Chattels:

HORSES

Black Gelding 7 yrs., about 1500 lbs.
Black Mare, 12 yrs., ab't 1500.
Gray Gelding, 5 yrs., ab't 1500.
Gray Gelding, 12 yrs., about 1300 lbs.
Sorrel Mare, 3 yrs., about 1300.
Black Mare, 3 yrs., about 1300.
Sorrel Mare 10 yrs., about 1400.
Bay Gelding, 4 yrs., about 1250 lbs.; Bay Mare, 5 yrs., about 1250 lbs.: Matched.

HARNESS

Set Breeching Harness. 3 Sets Plow Harness.
Number of Horse Collars.

CATTTLE

Roan Cow, 7 yrs., fresh. Red Cow, 7 yrs., fresh.
Red Cow, white-face, 6 yrs., fresh.
Red and White Cow, 10 yrs., due to freshen in May.
Red Heifer, due to freshen in May.
Red Cow, white-face, due to freshen soon
Pure Bred Shorthorn Bull, 2 yrs., from imported stock.
3 one-year-old Heifers. 5 Calves.

PIGS

9 Feeder Pigs, about 170 lbs. each. 8 Weaner Pigs.
Brood Sow, due to farrow soon. Brood Sow.

MACHINERY (in first class condition)

Hart-Parr Tractor, 18-36, all re-conditioned.
Cockshutt 3-bottom Plow. Spring Tooth Harrow.
Massey-Harris 21-ft. Disc Harrow. Deering Mower.
Van Brunt Seed Drill, 24-run, double-disc.
McCormick-Deering Binder, 7-ft. 2-wheel Trailer.
McCormick-Deering Gang Plow. 2 Wagons.
5-sections Diamond Harrows and Harrow Cart.
Set Sleighs, Massey-Harris. Triple Wagon Box.
Bulldog Fanning Mill. New Wood Water Tank.
Tank Heater. Well Pump, Pipes and Cylinder.
Pump Engine, McCormick-Deering. Pump House, 14x16.

HOUSEHOLD EFFECTS—Kitchen Range; Table and Chairs; Heater; Baby Crib; [illegible]

POULTRY—2 Turkey Hens. Gobbler.

300 Bushels Oats, 80% germination. Some Hay.

Also Forks, Shovels, Tools, Chains, Eveners, Neckyokes and other articles too numerous to mention.

TERMS, CASH --- All articles to be settled for before being removed.

Wallace Wade Auctioneer

GEO. WARREN. Clerk.

License No. 4-43-44.

Chapter 14

While father was exploring Armstrong, Spallumcheen and Vernon in the Sunny Okanagan, Patrick was having his fourth birthday and Anne decided to take him in a one-horse-open stoneboat ride out into the snow. Mother helped her hitch Percheron Pete, and she opened the gate for them and away they went towards Lornedale School. All the students were out for afternoon recess. Anne drove into the schoolyard for the students to see Patrick and to wish him a happy birthday. She got off the stoneboat, accidentally dropping the driving rein and Percheron Pete took off heading towards our home. Some students ran after the horse shouting, "Patrick let go! - roll off into the snow" and "Stop, stop, ho! Stop." The louder they yelled, the faster the horse ran and soon was in full gallop with one runner in the side road snow bank, and one runner on the road. Mother said she was putting Patrick's birthday cake into the oven when she heard a whole lot of yelling; and as she opened her kitchen door, she saw Percheron Pete standing beside the manure pile out on the field next to our garden patch. Patrick was walking towards her along the pathway from the horse barn looking like a little snowman. The students and Miss Nay, the teacher, arrived puffing and panting. Patrick told us that the horse stopped by the horse barn door and watched as he got himself out from under the snow and off the stoneboat; he then ran off to the manure pile; and that is where he stayed. The teacher and all her students stayed awhile and they all had some birthday cake.

When father returned he said he had made an agreement with a handshake with Mr. Berry of A. E. Berry Real Estate in Vernon; that he would go back to his farm and with the sale of his grain, livestock, and machinery he would make a deposit on the 18 acres of orchard with a brown cedar mansion on a hill overlooking a grassy meadow that went all the way to the shores of Swan Lake. Father always kept his promise.

An auction was arranged. Posters were printed and distributed throughout Innesfree, Ranfurly, Viking and were hand delivered to all the farmers. After the stooks were threshed and the grain hauled to the elevator in Ranfurly father scrubbed, greased, oiled and then painted all his machinery and with the Hart-Parr tractor were lined up from the garage and the pumphouse. He used saddle soap on the saddles, bridles, harnesses and halters, and he polished all the harness jingle bells and the buttons with silver polish and Brasso. Our cattle were brushed.

Edward and I went off to the pasture to bring in the horses. We carried with us a bucket with some oats in it and a rope. Our horses were in a summer fallow field together with our neighbours horses in the far north west end of our pasture. We were half way across the pasture when Percheron Pete saw us coming and came galloping towards us. As he was munching away at the oats in the bucket we carried, Edward helped me up onto his back; then the fun began. Percheron Pete knew what he was supposed to do. At full speed, he galloped back amongst all the horses, and he kicked with both hind legs, or with one, then with the other at the horses behind him. He would leap and bite at the horses in front of him; leaping, kicking, jumping as fast as his legs would go and all I could do was hang on, with both hands, for dear life, to the mane on the back of his neck. I flew up onto his neck to his ears, then back on his rump to his tail where I got whacked with his muddy tail. I swung from side to side over the horse's back like a pendulum do. The horses were rubbing and bumping close together in mud to their bellies. Each time I swung off the horse's back to one side of him or to the other, I would get pushed back up onto his back by a passing horse as they stampeded, leaped and kicked at each other through that soft muddy summer fallow spring thaw. Percheron Pete got the horses separated, and at a full gallop after our horses with mud flying everywhere, towards our pasture gate. After all the horses were through the gate and on their way home; I pulled at Percheron Pete's mane as hard as I could and I yelled "Ho!" to stop him. We were to wait for Edward; he was far behind us carrying the rope and the bucket with oats in it. Suddenly but very suddenly from a full gallop without warning Percheron Pete dug his fours into the ground and he stopped. Not expecting a sudden stop, I flew over his head to the ground below and that is where I stayed until Edward arrived. The horse was snorting, stomping, clawing at the ground and rearing on his hind legs above me. For as long as I sat on the ground before him, he did not leave, or run away. When Edward came, he threw the rope around the horses neck; we then led him home from the pasture gate along the fence. The horse was on one side of the fence, and we were on the other. Percheron Pete was behaving like a crazy horse and we did not trust him; he wanted us to let him go; he wanted to be let loose to be in the yard with all the other horses. I do not know why we hung on to him, but we did. Near the yard and because of the trees there, we let him go. With a twist of his tail and a nod of his head he was off. In the yard, he first went amongst all the horses as if he was inspecting or counting them; he then trotted over to father, nudged him with his nose and shook his head up and down then took a bow. Percheron Pete was telling father that it was he who brought the horses home.

"Leave the thinking to the horses - they have a bigger head" is true. Horses do know, and they do understand. Father always rewarded Percheron Pete for a task well done with an extra serving of oats. All the horses were brushed, their mane and tails trimmed and combed.

Everything was ready; sparkling, polished, brushed and clean. The day of the auction was here.

Percheron Pete was the first horse to be sold as all our neighbours wanted him. Then Duke and Beauty, the Clydesdales, they were a super special team though much too big to take in a boxcar. As I watched most of our farm animals leave one by one and two by two, the pigs, chickens, ducks, geese and the turkeys; all our farm machinery and our dog Rex; I could not quite understand what was really happening. Soon the auction was over leaving our yard looking lifeless, empty and sad.

Father kept five milk cows and a team of horses, Dolly and Dan. They were a bay, reddish-brown colour with black mane and tails. Their conformation was Morgan, though they were a smaller breed. Father also kept two pigs, some chickens and a pair of ducks.

Father drove to Ranfurly and sent the money to Mr. Barry in Vernon and he hired a boxcar for mid June when school would be out. I was completing grade eight, Edward grade five, and Henry grade three. Father also hired a man to travel in the boxcar.

One morning in June after the cows were milked, Henry and I were chosen to chase them on foot to the Ranfurly, stock yard. We each took with us a big long stick and father opened the gate for us out to the country road and away we went with our big sticks, chasing the cows. Father soon followed us with a wagon load but first he had to load the wagon. He would pick us up and have us ride with him he said.

Henry and I arrived at the Ranfurly stockyard tired and thirsty and the yardkeeper brought us a tea kettle of water to drink which we drank from the spout, and we waited.

Father arrived with Edward about two hours later. He said he thought we took a wrong turn, he did not expect us to move so fast on foot with the cows. He unloaded the wagon into the box car, and left the horses in the stock yard with our cows. He then hitched a ride with a neighbour back to our farm taking Henry with him. Edward and I were left at the stockyard in charge.

Father delivered mother, Anne, Henry, Patrick, Marshall and baby Joyce, in the car with all their suitcases, bottles, bags and diapers to Julianna's oldest daughter Eva's farm on the outskirts of Ranfurly where they would stay and wait for his return. When he returned he said his hired man would not travel in the boxcar as he was afraid father was sending him not to the other side of the Rockies but to the other side of the earth. Father loaded the cows and the team into the boxcar and the train slowly pulled away from the Ranfurly stockyard with father, Edward and me in it. We had with us beds, mattresses, pillows, furniture, dishes, pots and pans, saskatoon preserves and vegetables; sauerkraut, Sealcheeson, Kolbasa, cabbage rolls, pyrogys, cheese, homebaked bread and butter; and the cows, a team of horses, a crate of chickens, two pigs and a pair of ducks.

I did not know when I left our yard to chase the cows along our country road, that I would never again see my beloved farm, my home, without saying good-bye.

Fourteen years of building, farming, battling snowdrifts, sub zero weather, hale storms and tornadoes. Country life was over.

Father hired a boxcar, loaded it with farm equipment, animals and furniture and ventured out into the mountains, to the Okanagan Valley where his four sons and three daughters could get a higher education.

Hero, our black spaniel dog, was left on the farm as he was deaf and almost blind. The farm was the only home he knew and guarded with his life through the years. He fought skunks, porcupines, weasels, timber wolves, coyotes and foxes. Now he was old and could not be moved. A neighbour promised to feed and care for him.

Our Hero was home alone.

Colts in for aucton- Edward

Pump house, garage, stray dog

Cows in for aucton

Farm machinery - 1943

Henry behind colts
Edward

Patrick in front of cowbarn

Chicken house, Hart-Parr tractor and discs

Harvest

Threshing

Threshing gang

Winter on the farm.
Going sleighriding

Farm machines lined up for auction

Horses and horse barn

Lornedale school - May 1941

1952 - Horsebarn being taken down

1952 - House before being torn down

1993 - Anne where farm buildings and house used to be

1993 - Ranfurly - one elevator left standing
The stockyard our boxcar left from is no longer there

Chapter 15

As the train was slowly leaving the Ranfurly stockyard, my father was rearranging all our furniture which was stacked to the ceiling at one end of the boxcar. Bundles of hay and straw were stacked to the ceiling separating the livestock from the furniture and our food packed in oak barrels. Some food was packed in ice. The ice used was ice from our lidonia on the farm. Father was making a sleeping area, a little bed for each of us, and he set up a wash stand with a basin, soap and towels; and a potty. We had water for washing, water for the animals, and for drinking.

When the train stopped in Mundare, grandmother was there waiting to see us with a basket of food for us. The train stopped in Mundare for a long time; and while we were there I milked our five cows. Grandmother then took some of the milk into town and gave it to the people who needed it. Father raised these five cows from tiny calves, they were his chosen quality stock. He could not sell them at the auction, and here we were, in a boxcar, in a stockyard, with these five cows, waiting to be hauled away by a freight train, to a far away unknown land.

Freight Train

As the sun was slowly creeping over the horizon the next morning, we were approaching Edmonton. The boxcar had two half doors on the one side, the right side, a top half and a bottom half, and each half could be opened separately. On this calm warm sunny morning, my father opened both the half doors to let the sun shine in. When the train stopped I was milking a cow, father was standing beside me, and Edward was sitting up on top of the haystack near the ceiling watching when a policeman quietly approached.

The surprise of seeing a policeman startled us for a moment. Edward slid down the haystack to the other side amongst the furniture to hide, but father said, "Would you like some fresh milk? If you bring some containers, you may have some." The policeman smiled and said yes he would and then left. Before long, he returned, after he got his milk, he said he was curious; when he heard cattle lowing, pigs squealing, chickens cackling, a rooster crowing, horses thumping their hooves and ducks quacking, he had to come and see. He then thanked us for the milk, wished us a happy and safe journey and left. That evening, when the train was pulling into Edson, Edward and I stayed hidden as we now knew children were not allowed to travel in boxcars or freight cars.

Since the visit from the policeman in Edmonton, I milked the cows when the train was moving, bouncing, rocking and rolling. Father stood between the cow I was milking and the one beside her to keep them from bouncing together; from squashing me and from spilling the milk. I was the milkmaid all the way to our destination. I wore the pretty dresses my mother made for me before we left and my hair was a mass of curls, a permanent she made for me for the trip.

I do not know if my father slept a wink. He had to keep the horses calm day and night by talking to them and brushing them. Dolly and Dan had never been confined and tied to a small moving stall. Dolly was a quiet horse but Dan was clawing and thumping at the wall. He was spooked and was trying to climb the wall to get out. He caused a lot of trouble.

Father kept the floors of the boxcar clean. He put fresh straw down and fresh hay for them to eat and worked very hard while Edward and I slept in our little beds. With the sweet smell of fresh hay and the sound of animals near us I thought Edward and I were like the baby Jesus in the manger.

Edward fed grain to the pigs, chickens and the ducks and all the milk we could not use.

When the train stopped on our fourth night, the soft chug, chug, chug sound of the engine echoed and bounced around for a long time. The sounds were many; like an orchestra playing, the notes were wild and hollow. Prairie night sounds were very different; they were stable and solid. We were in the mountains in Jasper National Park

At dawn; we could not see the sunrise as the mountains were too high. Father opened the top half of the door for us to see the world around us and Edward and I climbed up onto that half door each holding on to the door frame and swinging our legs in the fresh mountain air; amazed and afraid of this wild and rugged wonder world we were seeing and we were singing:

When it's springtime in the Rockies
I'll be coming back for you
My little sweet heart of the prairie
with your bonnie eyes of blue
Once again, I'll say I love you
When the birds sing all the day
When its springtime in the Rockies
In the Rockies far away

We knew father was missing his wife Katharine, his sons Henry, Patrick and Marshall and daughters Anne and Joyce, who were waiting for his return somewhere on the Alberta Prairie without a home; like Gypsy's wandering around from place to place, and all because father's hired man refused to travel in the boxcar when it was time to go. From Ranfurly, mother and her brood traveled by bus to mother's old homestead near Hilliard for a few days. Her young brother Joe was still farming that old homestead with his wife Mary, daughters Arlene and Louise, and his son Gordon. Then, they travelled by bus to Mundare to be at grandmother's cottage to wait for father. July fourth, 1943 arrived and father had not come. That day baby Joyce had her first birthday at grandmother's cottage without father, Edward, and me. She must have been born under a wandering star.

A Wand'rin', Wand'rin', Star

I was born under a wand'rin' star,
I was born under a wand'rin' star.
Stayin' put can kill ya,
Standin' still's a curse,
To settle down can drive you mad
But movin' on is worse.
I was born under a wand'rin' star.

When I learned to talk,
the word they taught me was "good-bye",
that and "Where's my hat?"
Are all I'll need until I die.
Are all I'll need until I die.

Searchin' but for what I never will know.

I was born under a wand'rin' star,
I was born under a wand'rin' star.
I been to lots of places,
where I'd like to stay.
But when I like 'em best of all
is the day I move away.
I was born under a wand'rin' star,
A wand'rin', wand'rin', star.

Wandering has always been a big part of her life; and that is where it all began before she was one year old, on the Alberta prairie without a home.

After my mother's mother was widowed in the late autumn of 1934; she married a man who had a home in Mundare, and there, in Mundare, my mother and her brood were visiting both my grandmothers while father, Edward and I were in a boxcar, in the Rockies, travelling on.

The train was slowly winding around the mountains and we could see two engines pulling all the freight cars; and there were many. As the two engines went around a mountain then disappeared, the last car of this freight train, a caboose, was far behind us around another mountain and out of sight. We went through many tunnels. Some of the tunnels we went through were very long and very dark. In the tunnels, the only sounds we heard were the slow chug, chug, chug of the engines and the clanging, squealing, clattering of the wheels along the steel rails. The engines were powered by steam from burning coal in the engine firebox. The fumes from the coal smoke was black and thick, and the smell of the coal smoke was very unpleasant, and was choking us before we got to the other side and into daylight again. We travelled along long bridges from mountain to mountain. On looking down, we could not see the bridge, nor the rails, it seemed as if the train slowly stretched itself across from one mountain to the mountain on the other side of that deep, deep canyon below us; as a caterpillar does when it moves along from branch to branch of a plant shrub. Going over those railroad bridges was frightening, very frightening.

Around every mountain, through every tunnel, and before that long, slow stretch over bridges, from mountain to mountain, the train blew a sad eerie sounding note that bounced off the mountains for a long time; and down and through the canyons far below.

Nighttime came when the train came to a complete stop. The quiet, slow puff, puff, puff sounds coming from the two engines after the train stopped sounded as if the engines were gasping for breath after that long, hard haul pulling the freight cars through the rugged mountains; and here we were stopped for the engines to rest. There were no outside lights here where we were stopped; only a swinging, moving light along the railway tracks. A workman checking things out. We heard a sad, low, howl of a timber wolf, and the echo of the wolf howl and the repeated howling, sent a sudden fear through all of us in the boxcar. It sounded as if we were in a deep canyon and many wolves were all around us. The animals in our boxcar were restless all through that night; they were puffing, snorting and stomping as if they were trying to leap out of the boxcar. I could sense and feel that the timber wolves were circling our boxcar all that night. We did not sleep much during this, our last stop in the mountains. It was very cold here and spooky; and, we were exited knowing our journey was coming to an end. Here where we were stopped for the night; was Field, or it may have been Golden, British Columbia. It was too dark to see a name; and no one around to ask.

Six days on the road
and not many more miles to go.

One engine hooked on in the early morning, and we were on our way again. After we left that cold and frightful night in Field or Golden, we were no longer slowly chug, chug, chugging along. We were a-rattling, a-rolling and a-clanging along; it seemed as if we were a-flying. In no time, were we in Revelstoke, then on again. We arrived and were dropped off at the Vernon stockyard in the early afternoon. The engine that hauled us there to our destination, was a Canadian National Railway gasoline engine known as the "Galloping Goose".

Chapter 16

"An Apple Valley"

The next day, after six days travel we arrived in Vernon, my father's sunny "Apple Valley" with sparkling eyes, white teeth, and our faces black as coal; we landed in this beautiful Garden of Eden, towed in by a Galloping Goose.

"The Galloping Goose"

In 1910 the Canadian National Railway Company proposed a rail service from their main line in Kamloops through the Okanagan to Kelowna. Construction began at Kamloops in 1914, but the First World war, and lack of funding disrupted their plans. After the war, the federal government provided a subsidy of two million dollars and construction with a spur line to Lumby was completed in 1925. From Armstrong to Vernon, the Canadian National had running rights over the Canadian Pacific tracks avoiding two separate sets of tracks through the city. The "Galloping Goose" was an inter-urban gasoline engine. It first came to Vernon in 1926 and ceased to operate in the late 1950's.

When the train came to the Vernon stockyard we were excited, ready to face our new world. A moving truck was there to pick up and to deliver our food and furniture to our new home. Dolly and Dan, our horses, and the cows were sent down a ramp into the stockyard. The pigs, chickens and the ducks were left in the boxcar while father and the moving truck and driver left to unload our furniture in our new home. Edward and I stayed at the stockyard to guard our livestock. The Vernon stockyard had come to life as the horses galloped, jumped and leaped and kicked up their heels. The cows they swung their tails up over their backs and ran around after the horses. Before long people gathered and were sitting all around the fence watching, but Edward and I were not keen on talking to anyone. We jumped in amongst the horses and the cows and we rode, first the horses and then the cows; bareback as the Indians did; with our black faces, and white teeth smiling and my curly permanent full of coal dust. We kept on riding and entertaining the crowd until father and the moving truck returned.

The driver helped father load our wagon with everything that remained, and left. Father harnessed and hitched Dolly and Dan to the wagon and the three of us were on our way with father in the lead. Edward and I followed on foot herding the cows. We had to walk three miles north along the Kamloops highway to our new home. The highway was all curves, going past hills of orchard groves and farms. There was no straight section as our Alberta country roads. We stopped to admire the orchard farms with tree branches hung thick with little green apples. This was a different world. After a few more turns we arrived, at our mansion on a hill. Evening had come quickly that day; a shade of darkness was creeping in and our cows had to get milked, our food unpacked from the barrels and that night we had a very late supper. Our livestock was put into the barn for the night. And oh! a real bathtub; a free standing bathtub on four legs. A welcome relaxing treat after six days in a boxcar. No more round galvanized tubs to be brought in. We unpacked our sheets, pillows and blankets, made up our beds and slept. Father had worked very hard to get our home comfortable as best as it could be for Edward and me and tomorrow he was driving back to his sweetheart of the prairie; with her bonnie eyes of blue and all the rest of his munchkins awaiting his return. At dawn a car drove into our driveway and father with his suitcase in hand got in, and away they went and Edward and I were left alone in our new home.

This new home was very large. A wide curved staircase leading to the second floor, were four large bedrooms and a study or sitting area facing the front; with glass French doors leading out onto a view deck. On the main floor was the master bedroom, a full bath, living room and dining room with a granite rock fireplace; with rocks to the ceiling. At the front door was a large verandah the whole width of the house, supported by large square pillars. In the front yard were tall trees with bushy branches. Many little birds were in the branches of those trees chirping and singing all day long. A hedge of pink, white and red rhododendrons were amongst those trees, surrounded our front yard and rock walkway leading to the wide steps up onto the verandah and to the front door. The entrance room was all rosewood panelled with a walk in cloak room, sitting room, and the curved staircase leading to the second floor was at this front entrance. Our kitchen was very large too; with a family dining and breakfast area next to a bay window overlooking our apple orchard, and the barn and corral for our horses. Edward and I would sit at the breakfast table and watch the horses romp and play. That scene of seeing our horses was comforting and familiar; and kept us from getting too lonesome and of feeling all alone. To one side in our kitchen was a pantry with many shelves. Edward and I unpacked our mothers many preserves and we placed them onto the shelves in that pantry. There were wild strawberry, raspberry, saskatoon preserves and garden pickles of every kind; there were mustard pickles, sweet pickles, beet and horseradish pickles, dill pickles and more. In that little house on the prairie we left behind, mother worked very hard and many hours to keep our cupboards full. She canned many jars of chicken, whitefish, beef stew, tomatoes, corn, beets and more and here they were; all her Alberta preserves were here with us in this Apple Valley.

Our property extended from a hilltop down to the Swan Lake below. Edward and I often went down to the lake after we milked the cows, separated the milk and made butter, and we watched the swans gracefully swimming and guarding their young. This meadow that ran down to the lake was where our cows were kept during the day; and here in this meadow was a garden of planted strawberries. They were big; they were as big as golf balls and they were red and ripe. We picked them every day, and every day we ate fresh strawberries and cream and cornflakes. We walked the three miles to Vernon City to a Farmers Market Store once or twice a week; there we sold our butter, cream and farm fresh eggs and we bought cornflakes. Cornflakes was our special treat; and an ice-cream cone. Every time we walked to the Farmers Market Store, we bought an ice-cream cone, that was a must. We also sold cottage cheese to the Farmers Market Store, cheese we made from raw slightly sour cows milk; but that was not made often.

One early afternoon after it had rained and rained all night long, our yard to the barn and apple orchard was very muddy. Edward and I were in the yard barefoot with mud squishing in between our toes when a big fancy car drove into our yard, right up to where we were. A man stepped out; he was not a big man, he had a bald head and he was wearing white shoes and an expensive suit and tie. To us he said; "You are in a city now, not as you were on a farm; and here in a city you do not walk around barefoot. You must put your shoes on." He said, "I knew your father when he was a little boy, and I know your Grandmother; your father's mother. When will he be back?" We said we did not know; he then left and we never saw him again. He did know my father, though they did not meet often.

Our apple orchard had tall grass and things growing amongst the apple trees. One day Edward and I were going to cut that grass and things down with a sickle. Then we decided to let the cows into the orchard to eat up that grass and to flatten the weeds and things. The cows went far into the orchard and they kept on going and going. We did not know that those things with the big wide leaves growing everywhere were burdocks. We tried to go into the orchard after the cows and those burdocks grabbed onto our clothes, and held on. When we tried to pull them off they pulled apart, not off. By the time we made it through that burdock jungle to our cows, our clothes were all stuck with sticky burdock, and by the time we reached the cows, their legs, bellies and their tails were thick full of sticky burdock. We spent hours picking and pulling and brushing them off.

After father left us in our new home alone, Edward and I dragged one bed out into our living-dining room and we set it up from where we could see the kitchen, the front door, and the upstairs. And we brought a long handled axe up from the basement and we placed it under that bed; and that is where we slept; both together in the one bed. We did not want to be closed off into one room, and not know what was happening throughout the night in that whole big house. To this day, I have no idea what we would have done with that axe, or why we put it there.

In the orchard, we cut down most of the burdocks with a sickle and an axe; we then let the cows back in to eat up the grass.

Along with the grass, they ate a lot of the little green apples off the ground and they got sick. We did not know why they got sick, but we kept them in the barn for a few days and they got better. We then sent them down to the lake's edge into the meadow.

Edward was eleven and I was thirteen and we were trained from very young to be brave, responsible and to do a job right. Father was a perfectionist and so were we. A job, any job, had to be done absolutely right or not to be done at all. Cleanliness was top of our list. Cleanliness and sterilization was of the utmost importance. We sterilized our crockery containers for making cottage cheese in, and we sterilized all our utensils for milk, cream, and for making and storing butter. My mother taught my sister Anne, and she taught me how to be a woman, to cook, bake, to do laundry, clean house, to sew, embroider, crochet., to care for and to look after babies. Most of all, to give respect to and tender loving care to the master of the house, the husband, the father of our children; our provider. Those were the rules families lived by when Canada was being born; when the first settlers came, and during the depression years. Those were hard times, but happy times.

During the Pioneer years; and during the depression years, people had very little or no money, but they were strong, they were together and they were a happy family. The war years, and the depression years, that families struggled to get through, were being left behind them and they were beginning to see the sunshine, with clear blue skies ahead and they were believing:

There'll be bluebirds over
The white cliffs of Dover
Tomorrow
Just you wait and see.

Then some women that were not sure if they were man or woman disrupted our future by demanding equality for men and women. 17th of April, 1982 the Queen Elizabeth of England came to Canada and together with Canada's then Liberal Prime Minister, Pierre Trudeau, they signed the Constitution Act for equality:

Did they really know what they were doing?

4. Commencement (s. 58)

Section 58 of the Constitution Act, 1982 provides that the Act is to come into force on a day to be fixed by proclamation. That proclamation was issued by the Queen, who came to Canada for the purpose, at a ceremony in Ottawa on April 17, 1982; and the proclamation fixed April 17, 1982 as the day upon which the Constitution Act, 1982 was to come into force.[75] The Charter of Rights accordingly came into force on that day, and operates only prospectively from that day.[76]

Federal and Provincial governments were given three years - until 17th of April, 1985 to make their laws conform to the equality clause. Most provinces began allowing women to use their maiden names and to give those names to their children. Some provinces started to let spouses sue one another for damages in divorce cases. And they no longer allowed a wife's "misconduct" to interfere with alimony and family maintenance payments.

Women began battering their husbands and getting away with it.

Since the first government welfare cheque, during the 1970's, was issued to a woman, many women began leaving their husbands and their children, for the money. And Country Canada Chaos began, and that chaos spread throughout the country like wildfire. Journalist Doris Anderson became popular with some women with her personal beliefs and with her published books on "Woman's Rights" and a "Woman's Liberation Movement" began to fester. Envy, resentment, and hate began to fester within families. Women became demanding, to become priests, as Parliamentarians, lawyers, judges, sports reporters, football and hockey players, long-haul truck drivers, construction workers and so on, anything but being a wife and a mother, a treasured woman, as she was put on this earth to be.

less equal than **MEN**

Passage of the Charter of Rights and Freedoms does not mean that the fight for the equality of women is over; in fact, the battle has only just begun

"No woman, idiot, lunatic or child shall vote." That was the law when this country was new. At that time, Canadian women had no part in choosing the government representatives who made the laws they had to live under. With no voting rights, women had little hope of gaining equality under the law.

Women also want to have the right to enter and to be a part of a "Man's Private Club". What will be next? - To enter a man's washroom, dressing room and to stand up to piss?? Have women stopped to think of what fools they are?? What do they want?? It is not the women that are content at being women that are causing chaos, it is the castrated women that are making the world go crazy; and the brainless idiots are following like sheep.

Then came government interference; government rules and the law, to take children away from their parents, to foster homes, or to give complete and full control to a mother, without any rights to a father to raise a child; a mother who has no idea what the responsibility of childraising is all about. When government authorities began taking little children away from their parents; they were breaking those children's spirits, their emotions, their heartfelt love for living. Those little children who are young adults in the '90's, twelve-to seventeen year olds, their family togetherness, family security and belonging was shattered then and is forever shattered and the reason why most of the young generation in the '90's, are lost, are on drugs, and are committing suicide, and why there are more and more homosexuals, because there are no more fathers and feminine caring mothers in a home, but stepfathers or boyfriends, two men or two women, since the charter of equal rights for men and women was signed.

How can a little girl; a two, three, nine, or thirteen year old, climb up into a mans arms; a man who is not her father and feel protected, secure, safe and loved, which is the most important part of a little girl's development, in becoming an honest to goodness woman. Since the equal rights act was signed, there are more family breakups. Including the breakups in the Queen's own family - the Royal Family. And

Pierre Trudeau; - all the University doctorate degrees he allows his sons to get will never ever replace their growing up with a mother and a father together in a family home.

What is forever for?

My brother Henry said; and I quote - He said; "A marriage, your spouse and your child are equals. You do not discard a marriage, a spouse or a child: They are forever." And he was right; he was always right. But no one listened. My brother Henry said it - Rafe Vanhoy put those words into music and song - But nobody listened.

What is forever for?
I've been looking at people
and how they change with the times
and lately all I've been seeing
are people throwing love away
and loosing their minds.

So what's the glory in living
Doesn't anybody ever stay together anymore?
And if love never lasts forever, tell me:
What's forever for?

Maybe it's me that's gone crazy,
'Cause I can't understand why;
All these lovers keep hurting each other,
When good love is so hard to come by.

Since the 1980's - there are no more forever families with fathers and mothers in full control of their children: No more men and women - they are species.

I am fortunate that I was born and raised during the

depression years with my own mother and father: We had family togetherness, real value, belonging, compassion. Had my father moved us, his family from our Alberta farm to our new home in the sunny Okanagan the year 1993 rather than 1943, and left an eleven and a thirteen year old alone on an orchard farm for several weeks, we, Edward and I would have been taken off that farm by some government authorities and placed in a foster home with foster parents, shattering our lives forever. Government interference's into family affairs should never have become law. My father was the best; as a husband;,and as a father. He trained us well; he trained us to be brave and to survive through most difficult times as he had to as a little boy and through all his developing years. He loved us, his family, and he cared; with his whole heart and soul he cared. No one, but no one in this whole wide world could have ever replaced him. His love, his caring, his kindness and sincerity I will forever remember. He was a real man; he was the greatest husband to his Katharine and father to his sons and daughters.

The years with father were happy years. And I did not get to tell him that before he died.

I was not there.

Canada
The Institutions of Our Federal Government

As long as humans have been organized in societies there have been parliaments. Of course, the first parliament didn't look much like ours today, a group of elders might gather around the campfire at night to decide on weighty issues affecting the tribe; where to hunt the next day, what to do if the village waterhole dried up.

Generally, there wouldn't gave been much democracy in evidence. Elders were likely to be elected. Mostly, they held the power because they were the tribes oldest men; women and younger males didn't have much say in matters affecting them.

England is where the movement toward parliamentary government began. The Monarch had total control. The kings and queens of England had assembled councils of barons to advise them.

The barons were thinking only of themselves; they were not interested in giving the masses a voice in the affairs of the state. However, The Magna Carta of 1215 started a process of democratizing society that continues today.

Canada inherited its parliamentary traditions from England, and Ottawa was selected as the national capital by Queen Victoria in 1857.

The word parliament is derived from the French verb parler, to talk.

The Institution of Our Federal Government

By the Constitution Act, 1867, "the executive government of and over Canada is declared to continue and be vested in the

Queen." She acts ordinarily through the Governor General, whom she appoints on the advice of the Canadian prime minister. The Governor General normally holds office for five years. Though the tenure may be extended for a year or so.

Parliament consists of the Queen, the Senate and the House

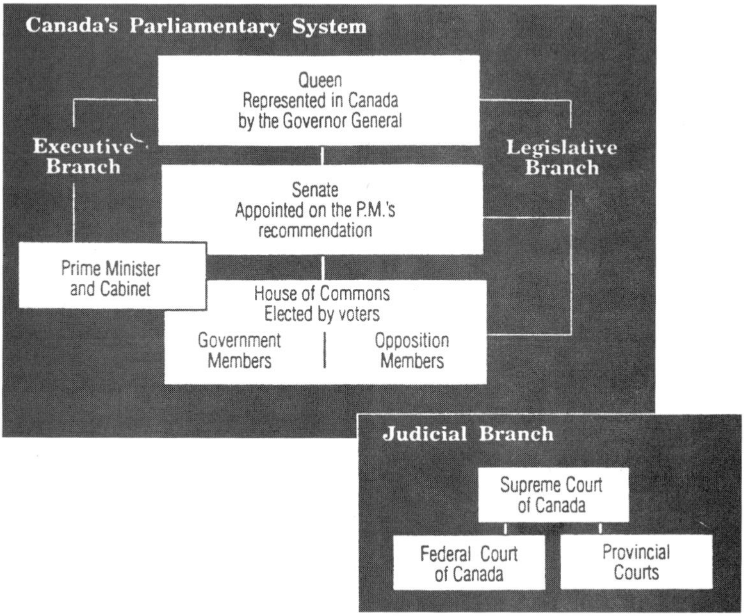

of Commons.

The Senate

The Senate has 104 members: 24 from the Maritime provinces (10 from Nova Scotia, 10 from New Brunswick, 4 from Prince Edward Island); 24 from Quebec; 24 from Ontario; 24 from western provinces (six each from Manitoba, Saskatchewan, Alberta and British Columbia); six from Newfoundland; and one each from the Yukon Territory and the Northwest Territories. There is provision also for four or eight extra senators, one - or two - each from the Maritime provinces, from Quebec, from Ontario and from the West. This was used in 1990 to ensure passage of the controversial

G.S.T. legislation.

The senators are appointed by the Governor General on the recommendation of the prime minister. They hold office until age 75 unless they miss two consecutive sessions of Parliament. Until 1965, they held office for life, and the few remaining senators appointed before that date retain their seats.

The Senate can initiate any bills except bills providing for the expenditure of public money or imposing taxes. It can amend, or reject, any bill whatsoever. It can reject any bill as often as it sees fit. No bill can become law unless it has been passed by the Senate.

The House of Commons is the major law-making body.

It has 301 members

Area	Seats
Ontario	103
Quebec	75
British Columbia	34
Alberta	26
Manitoba	14
Saskatchewan	14
Nova Scotia	11
New Brunswick	10
Newfoundland and Labrador	7
Prince Edward Island	4
Northwest Territories	2
Yukon Territory	1
TOTAL	**301**

Unbalanced Power

Canada has prospered under its current systems of parliamentary government, but there are many experts who believe it's time for a change.

Individual MP's are powerless and can't represent the wishes of the constituents effectively;

The regions outside central Canada do not have an adequate voice in Ottawa;

The Prime Minister, his advisers, and an inner circle of Cabinet Ministers have too much power; and

The Constitution has dominated the time of our leaders at the expense of more pressing issues for the last 40 years. For most Canadians who pay any attention, Question Period is the only part of the parliamentary process they see; unfortunately, it is probably Parliaments' least important activity and the one that shows politicians in their worst light.

1965

Canada adopts its own flag - the Maple Leaf.

17, April 1982

The Royal Proclamation was signed by Her Majesty the Queen, Prime Minister Pierre Trudeau, Justice Minister Jean Chretien and Registrar-General Andre Ouellet bringing into force Canada's Constitution.

Political Parties

Our system could not work without political parties. Our major exisitng federal parties - Progressive Conservative, Liberal and New Democratic - were not created by any law though they are now recognized by the law. We the people, have created them ourselves. They are voluntary associations of people who hold broadly similar opinions on public questions.

1992

The Beaudoin-Dobbie report issued by the committee for a Renewed Canada recommends an elected Senate and equality of "people of the regions" entrechment of the inherent right of Native self-government, and power-sharing deals between the federal and provincial governments. The quebec National Assembly rejects the report.

Aboriginal leaders receive an invitation to become " full participants" in constitutional negotiations. A federally sponsored conference of Aboriginals ends with no clear definition of self-government or how to implement it.

The Constitution Acts

First ministers, and territorial and aboriginal negotiators, reach the Charlottetown Accord which includes a reformed Senate, native self-government and greater provincial power.

The Charlottetown Accord is rejected in a national referendum.

Canadian prime ministers since 1867

1.	Rt. Hon. Sir John A. Macdonald	Liberal-Conservative	July 1, 1867 to Nov. 5, 1873
2.	Hon. Alexander Mackenzie*	Liberal	Nov. 7, 1873 to Oct. 8. 1878
3.	Rt. Hon. Sir John A. Macdonald	Libèral-Conservative	Oct. 17, 1878 to June 6. 1891
4.	Hon. Sir John J.C. Abbott*	Liberal-Conservative	June 16, 1891 to Nov. 24, 1892
5.	Rt. Hon. Sir John S.D. Thompson	Liberal-Conservative	Dec. 5, 1892 to Dec. 12, 1894
6.	Hon. Sir Mackenzie Bowell*	Conservative	Dec. 21, 1894 to April 27. 1896
7.	Rt. Hon. Sir Charles Tupper* (Baronet)	Conservative	May 1. 1896 to July 8. 1896
8.	Rt. Hon. Sir Wilfrid Laurier	Liberal	July 11, 1896 to Oct. 6. 1911
9.	Rt. Hon. Sir Robert L. Borden	Conservative	Oct. 10, 1911 to Oct. 12. 1917
10.	Rt. Hon. Sir Robert L. Borden	Conservative**	Oct. 12, 1917 to July 10. 1920
11.	Rt. Hon. Arthur Meighen	Conservative	July 10, 1920 to Dec. 29. 1921
12.	Rt. Hon. William Lyon Mackenzie King	Liberal	Dec. 29, 1921 to June 28, 1926
13.	Rt. Hon. Arthur Meighen	Conservative	June 29, 1926 to Sept. 25.1926
14.	Rt. Hon. William Lyon Mackenzie King	Liberal	Sept. 25, 1926 to Aug. 7. 1930
15.	Rt. Hon. Richard Bedford Bennett (became Viscount Bennett, 1941)	Conservative	Aug. 7, 1930 to Oct. 23. 1935
16.	Rt. Hon. William Lyon Mackenzie King	Liberal	Oct. 23, 1935 to Nov. 15. 1948
17.	Rt. Hon. Louis Stephen St. Laurent	Liberal	Nov. 15, 1948 to June 21, 1957
18.	Rt. Hon. John G. Diefenbaker	Progressive Conservative	June 21, 1957 to Apr. 22. 1963
19.	Rt. Hon. Lester B. Pearson	Liberal	Apr. 22, 1963 to Apr. 20, 1968
20.	Rt. Hon. Pierre Elliott Trudeau	Liberal	Apr. 20, 1968 to June 4, 1979
21.	Rt. Hon. Charles Joseph Clark	Progressive Conservative	June 4, 1979 to March 3, 1980
22.	Rt. Hon. Pierre Elliott Trudeau	Liberal	March 3, 1980 to June 30. 1984
23.	Rt. Hon. John Napier Turner	Liberal	June 30, 1984 to Sept. 17, 1984
24.	Rt. Hon. Martin Brian Mulroney	Progressive Conservative	Sept. 17, 1984

** Prior to 1968, "Right Honourable" was accorded only to Prime Ministers who had been sworn into the Privy Council for the U.K. Prime Ministers Mackenzie, Abbott and Bowell were only members of the Canadian Privy Council and Prime Minister Tupper became a U.K. Privy councillor after his term as Canada's Prime Minister.*

*** During his second period in office, Prime Minister Borden headed a coalition government.*

History of the National Anthem

- "O Canada" was proclaimed Canada's national anthem on July 1, 1980, 100 years after it was first sung on June 24, 1880. The music was composed by Calixa Lavallie, a well known composer of operettas, symphony and songs. He was a pianist and organist and he was a teacher, a "Canadian errant" born in Verchères, Lower Canada on December 28, 1842. He died on January 21,1891 at the age of 49. French lyrics to accompany the music were written by Sir Adolphe-Basile Routhier (1839 - 1920) born in Saint Placide (Lower Canada). He was appointed a judge in 1873. The song gained steadily in popularity. Many English versions have appeared over the years. The version on which the official English lyrics are based was written in1908 by Mr. Justice Robert Stanley Weir (1856 - 1926) born in Hamilton, in what was then Upper Canada. This is the version which was published in an official form for the Diamond Jubilee of Confederation in 1927, and has since been generally accepted in English speaking Canada. His fame as a writer won him election as a fellow of the Royal Society which Routhier had helped found.• French words, the first stanza is now generally the only one that is sung.

Sergeant — North-West Mounted Police 1874   Sergent — Police à cheval du Nord-Ouest 1874

O Canada! terre de nos aïeux,
Ton front est ceint de fleurons glorieux.
Car ton bras sait porter lépée,
Il sait porter la croix;
Ton histoire est une épopée
Des plus brillants exploits;
Et ta valeur, de foi trempée,
Protégera nos foyers et nos droits.

Sergeant — Royal Canadian Mounted Po

We st

O

O Canada! Our home an
True patriot-love in all thy
With glowing hearts we
The True North, stro
And stand on guard
We stand on guai

Chapter 17

O Canada we stand on guard for thee. Who will stand on guard?? - and what for? First the native Indian children were taken away from their birth homes and families, and since, white Canadian children are being taken away from their homes and birth parents. With spirits broken they do not have the emotion, the strength nor the desire to sing - So! Our Federal Liberal Prime Minister, Jean Chretien spent **Fifteen Million, Five Hundred** and **Ten Thousand Dollars** and **No Sense.** of federal funds on flags for them to swing.

My heart cries for you.

Our national anthem, the words and the true meaning of, has been changed, and today is seldom sung or heard.

School days during the 30's and 40's, every day began with all the students singing "O Canada our home and native land"; with true emotion and strength they sang that made the walls of the schools vibrate. After the singing was done; they were taught the 3-R's - reading, writing and arithmetic. Today there is no more singing of our national anthem. Students are being taught to accept, to respect and be kind to homosexuals. Do homosexuals wear a patch on their foreheads for all to see? And why would they want to advertise their private preference. The visible disabilities suffer so much more abuse and do not demand acceptance or compassion. What should be taught in schools today, is "What forever is for". A job requires devotion, reliability, responsibility and trust. Those same qualities plus much love are required in a marriage, a home and children - forever. What is being taught in our schools today is not what out ancestors, our Canadian pioneers were building Canada for. Today and every day, in every school students should be reading, over, and over, and over again about our "Pioneer Years", as they would read a Bible so they would know how Canada was built, who built Canada, and the foundation that was set by those people. How could we have allowed the Native peoples of our Canada who welcomed us, the whites, as they named us, into their country. How could we have such a low respect and acceptance of our Native peoples, and give so much power and control to the new arrivals who are not familiar with, nor do they know our way of life, our people and our Canada. And why should old Canadians be taxed to pay for the new arrivals to learn to speak English when that should be a private class and should be paid for privately, as music and song. I studied "Political Economy" in the Ukrainian language as a teenager in Vernon, British Columbia. A special class taught by Professor Mike Terebenetz; to be able to understand the Slav peoples, the Ukrainians, the Polish, Czechoslovakians, Yugoslavians, Austrians, Hungarians, Bulgarians, the Jewish and the Germans, of why they came to Canada, their desires to stay and to build this country Canada. I wanted to know of their labours and of their life with the Native Indians. The pioneers of 1895 - 1914 did not look for power nor for handouts. And! They did not force their style and their way of life onto anyone. They were Canadians and they worked, and they worked hard to build this country, our Canada - "with patriot love" "and with glowing hearts" to see "our True North strong and free". And "O Canada they stood on guard for thee". Professor Mike Terebenetz returned from the second world war so battered up he was barely alive, though he returned, his willpower and determination to see his people,his Canada, his pioneers before he dies. He was a most kindly, learned man of all time. He had no family of his own; just his pioneers. After the second world war ended in1945; no one expected that the East Indians and the Vietnamese would be allowed to change the appearance of our Canada to look like Indo-China.

Canada is a unique country, and should continue to be. The native peoples are the Emblem the Logo. The people which is Canada forever: and the Royal Canadian Mounted Police are to protect that image with their red coats, navy riding pants, Stetson hats and high brown boots. That is the Canada they stood on guard for. That is the Canada they built and died for; and that Canada foundation is now crumbling. Canada today is becoming the story of

Humpty Dumpty who
sat on a wall
Humpty Dumpty
had a great fall
All the King's horses

and all the King's men
Could not put Humpty Dumpty
back together again.

We must wake up and put Canada back together again before it's too late. All the turbans and all other country garb do not belong on the streets of our Canada. Fine theatres were built throughout our Canada for costumes, song and dance of other nations. That is where costumes belong; in the theatres not on Canadian streets. The Native Canadians are the only nation that should be on the streets of all Canada showing their nations dress, song and dance because they are Canada. "KANATA" which means "village dwellers" by the Huron Iroquois "KANATA - village community" is Iroquoian.

White Canadian "hen pecked" men; wake up; you are the strong; be the proud and save our Canada. As our Native people say, "Man! you are the sun, you are the moon, you are the sky. You are who nurtures mother earth." And I say: "Woman! Come back down to earth." "Woman's Equal Rights" if allowed to continue, will complicate all of life; family and human values will no longer exist; and this must stop. My father was the master of our home and of our destinations. He made all the decisions, and he alone was in full control of all his money through the depression years and beyond; had he not been, we all would have perished. How very lucky we all were that he was strong willed and that he guided us all; his family, into a responsible life.

Food banks, welfare handouts, a "Manna from Heaven" will not make today's immigrants and all Canadian youths from broken homes become responsible citizens.

Liberated women are the cause of youth destruction.

Chapter 18

21st July 1943; - a bright warm sunny Wednesday afternoon with a few cloudlettes drifting by high up in the sky. Edward and I had been to the Farmer's Market Store on the Monday, the 19th. That day we sold our cream, butter, farm fresh eggs and some cottage cheese. In our pantry was a quart size jar full of money we stored from our sales, for father when he returns. Our work was all done; the cows were down in the meadow by the lake's edge and that day we sent the horses up into the hills above our house and garden. The garden was a two acre patch of beefsteak tomatoes and beans growing for Bulman's Cannery. In the big trees beside our house, the many little birds were chirping and singing, blue jays flying through the orchard were screeching and complaining. Robins in the yard were hopping around and listening to the ground beneath them, then, when found, would start pulling at a big fat earthworm till they tumbled over backwards when the worm came out of the ground. With worm in beak they would then fly away. Edward and I were climbing trees; our apples trees. The branches on apple trees were easy to climb; they were thick through and many of them right to the top. The trees were twelve to fourteen and some were sixteen feet high. Edward and I chose a favourite tree each in a different row, and we climbed up to the top. As we were sitting there amongst the branches loaded with little apples, we were looking up into the sky and guessing the cloudlet formations. Some cloudlets looked like people's faces; some looked like little lambs. And we found a variety of animals, birds, bees, cotton balls and some looked like snowmen.

From the tree tops we could see the afternoon train on the other side of Swan lake chugging along towards the Vernon station. As we were watching that train and remembering our six day journey in a boxcar; a gun metal grey Plymouth car turned into our driveway. It was father; he had returned and brought with him our mother, and the rest of our family. Edward and I almost flew down from the tops of those trees we were on to the ground below with excitement; and came running out of that orchard to greet our family. What joy that was, a true sense of security, contentment, and wholesomeness. Family togetherness is real strength and encouragement. Being separated since we left the farm when Henry and I chased the five cows to the Ranfurly stockyard seemed like a long time, and here we were, all together again and in father's "Sunny Okanagan" his "Apple Valley" home.

While he was away; father said he took the family to see our farm - one more time. He said it all looked very empty though they all spent the day there relaxing and remembering. Mother's rose garden was in full bloom and the aroma of all the different varieties of beautiful roses filled the air. Here they found peace, quiet and contentment they didn't realize when this beautiful 320 acres of land, this dwelling, was home, and much work had to be done. While the little ones with Anne in charge were running around the yard, through the barns and out to the garden patch on the field where they picked a whole lot of big thick rhubarb sticks; father and mother stroll along hand-in-hand reminiscing of their days gone by, of the hard times, the work and of their babies that were born here. They pumped some water from the well before they returned to their little log house where they made tea. The old stove was still standing in the kitchen with a coal oil lamp sitting on top of it; and when Anne returned from the garden with Joyce, Marshall, Patrick and Henry, they all had a picnic lunch out in mother's rose garden.

Father said he rented his Alberta 320 acres with all the buildings to a family by the name of Petruk from Edmonton. That they would cultivate this land and grow grain for the next five years. Father also took his family to see his three sisters, Maria, Julianna and Paulina and mother's family, John who lived near Viking, her sister Tillie in Lloydminster, Saskatchewan and Joe near Hilliard on mother's old homestead. They stayed in Mandare a few days with the grandmothers before starting their drive through the mountains. When they arrived in Banff, they stayed at the Banff Springs Hotel for three days to relax in the Banff sulphur hot pools before driving the Big Bend Highway and on to their new home. Mother said the highway from Banff on, and through the Big Bend, was very narrow. It was frightening and she did not enjoy the drive or relax till after they left Revelstoke and the Stepney Valley was in sight. Mother brought with her a beautiful big bouquet of her roses from her rose garden. Most of the roses were dry; though they were her roses. They also brought a bundle of thick juicy rhubarb sticks Anne and the little ones picked on our farm garden. Before mother went in to see her new house, she ran out to the garden, then through the garden gate and up and over the hill where our team of horses were let for that day. When she returned with disappointment in her eyes, she said she wanted to see what was on the other side of that mountain our house was on, and all she could see were more and more mountains.

Mother loved the house, because it was big and had plenty of room for us all, though the mountains made her feel closed in. She could not see and enjoy the sunsets or the sunrise. She truly was "a girl of the prairie, with eyes of blue and a heart so true". For many years she missed her Alberta early morning sunrises then to sit on the verandah late in the evenings to watch the sun set. There was none of that here. Before we got reacquainted with each other and with our new way of life in the Okanagan, work had begun and there was no more time for remembering and missing our Alberta prairie home. Pole beans and tomatoes were ready for picking and we all were out in the garden picking and picking all day long and filling many crates. Then we'd load the crates onto the wagon and father hauled them away to the Bulman's Cannery. After Bulman's got their quota off our acres, we all helped to preserve the tomatoes and the beans that were left. Tomatoes; more and more tomatoes. Father kept buying jars and we picked; then put ten to twelve tomatoes into a cheese cloth, dip them into boiling water then into cold water. Slip the outer skin off and stuff them into sterilized jars. The jars filled with tomatoes with lids on, would then be placed into a boiler on the stove and boiled for three hours. That year; the summer of '43 mother preserved 300 jars of tomatoes, 250 jars of beans and apple sauce we lost count of. Every apple that fell on the ground off the trees we could not sell; we made into applesauce. We had applesauce on pancakes, applesauce on toast, applesauce for dessert with whipping cream every day. And we made applesauce and peanut butter sandwiches for school. Not one little apple, tomatoe or bean was wasted. The Pioneer years and the Depression years taught people to be thrifty to survive, and that survival training stayed within our families from generation to generation.

Life in this Paradise valley kept us all working day and night through the summer. There was no summer holiday for us, as we had in Alberta mid summer while waiting for our grain to ripen. Here we had three varieties of apples to pick. First were the Snows; they were an early apple. Inside they were snow white, with beet red skins. The MacIntosh were next to be picked and the Delicious apples were picked late in the autumn when the snow started to fall.

We had to learn a special way to pick apples. We carried thc twelve to fourteen and sixteen foot ladders to a tree, set it up and start picking off the lower branches first. The apple was to be cradled in the palm of our hand; then with a twist of a wrist upward, to break the apple stem away from the tree; to avoid bruising, the apple was not to be held with fingers. And we carried a big galvanized metal bucket over our shoulder with a canvas drawstring bottom which was released to empty the apples into a wooden apple box.

September arrived and there was still very much work to be done in our garden and the orchard. We also had apricots, peaches and Italian prune plums to pick. But our school days, and golden rule days were here. A school bus stopped at our front gate and Edward, Henry, Anne and I boarded that bus and were hauled away to a school and amongst students we did not know. Edward and Henry were delivered to an Elementary School and Anne and I were dropped off at the Polson Park High school. I tried to shrug off the nervousness I felt within me in hope, and in believing, that I do not know the students; and they do not know me. So I put on a brave front and up the steps to the front door I went; when I heard whispers behind me saying: "That is the girl that came to Vernon in a boxcar and with a black face and fuzzy hair she rode cows and horses in the stockyard". I wanted to disappear but that didn't happen. I was embarrassed and did not believe I would be recognized. Mr. Seaton was my homeroom teacher, a compassionate man he was; he approached me at the entrance door, welcomed me to the school, then introduced me to the class. That whole school term from September on, was not easy for me to adjust to and I failed my grade 9 in June. This was a most difficult time for us all; and I believe that my mother and father were feeling that they did not belong in this Paradise Valley.

A Mansion on the Hill

I've waited all through the years, Love
To give you a heart true and real
'Cause I know you're living in sorrow
In your loveless Mansion on the Hill
The light shines bright from your window
The trees stand so silent and still
I know you're alone with your pride Dear
In your loveless Mansion on the Hill

Spring of '44 came early in this sunny Okanagan Valley; much earlier than it did for us in Alberta. Here in this Paradise Valley, tomatoes and beans had to be planted and the apples had to be thinned out long before school was out for the summer. After clusters of blossoms fell off the trees and little apples were the size of small crabapples, they had to be thinned out, to make room for the chosen apples left, to grow a bigger size. One finger was used to snap the little apples out of the cluster; the third finger. Mother was out in the garden planting tomatoes, with Patrick, Marshall and Joyce at her side. They were helping her. Father let the cows into the orchard that morning and was thinning apples, when he noticed one cow, the yellow-cream coloured one with white patches, we called Buttercup, up on a rock cliff at the far end of the orchard. He was high up on the orchard ladder and immediately got down and ran through the orchard to the rock cliff to bring the cow back down. He was wearing leather soled boots he bought at an auction sale and he climbed a narrow footing along a barbed wire fence to the timorous cow not able to turn around to go back down. He tried to get her to go backwards and he slipped on that rock cliff and went tumbling down 10 feet to the ground below on his left leg and side. He ripped his pants and his leg was torn to the bone with his knee cap hanging out. The bleeding was so severe; had mother not been out in the garden to hear his calls, he could have bled to death. Mother went running to him; she ripped her petticoat, tied his wounds and helped him into the house where they cleaned his wounds with iodine. I do not know how they did it, I was at school. Father would not go to a doctor, he was like his mother. She, my grandmother was a young widow with no money. And he, my father was not going to spend money on a doctors services when he could doctor himself. His injuries were very bad and serious. Always, I was my father's "right-hand-helper" he said, and I was not the kind of child who had to be asked or told what jobs were to be done. Immediately I took over for my father all the work he could not do, and because I was willing, I was able, and because I was brave. I wanted to help, to do all I could do for my family.

Anne was a student who passed her grades, and she did milk a cow if the cow did not move a leg or swing her tail. That spring we also had a litter of little pigs to feed morning and evening. Edward and Henry fed grain to the chickens and gathered the eggs. The ducks were left outdoors waddling around in the yard and garden gathering slugs.

The day of my father's injuries; we left the cows in the yard, with access to the orchard for overnight. In the morning, when we went out to milk them, three of the cows were in front of the barn door on their backs with their legs and udders up in the air blown up like balloons. They were bloated and they were dead. A fourth cow was there too, beside the barn door gasping for air. And the fifth cow was standing near the apple orchard looking very sad. Edward and I ran to a neighbour who had a telephone and we called a veterinary doctor who came and punctured the gasping cow with a sharp object to her side to let the air out of her; and she lived. The veterinary doctor said we were not to feed apples to cows. That; we did not know. If cows can eat corn, why not apples? We were left with two cows.

Father could have had cracked or broken ribs from his fall; and he could have been bleeding internally. He did not say how he felt and he did not complain. How he stitched or bound his torn leg and knee to mend, is miraculous. His leg and knee did mend and he was able to walk again. Some years later; he did say that he should have seen a doctor. During his convalescence; he did not sit idle; he always found work he could do and during his spare time he handmade wire coat hangers for our clothes closets while he was planning a different way of life for us here in his Apple Valley.

School was out for the summer; and that is how our first year in this sunny Okanagan Valley had come to end.

Chapter 19

Vernon, British Columbia

Vernon is situated in a pass commanding and making tributary to it the whole of the famous Okanagan Valley and its adjacent valleys, the grazing country of the lower Nicola and Similkameen and the placer diggings of Cherry Creek and Siwash Creek. It is impossible to get to any of the above places without first going through Vernon.

Vernon was founded in 1867, because it is a natural distribution point. It prospered because early settlers had a vision of its future and worked hard to achieve it. Many settlers came to take advantage of a pre-emption plan under which 160 acres of crown land could be had virtually free. Vernon's pioneer period lasted a mere thirty years 1858 - 1885. It ended with the completion of the trans-Canada railway which in 1885, passed within fifty miles of Vernon, with a station at Sicamous. In 1891 the Shuswap and Okanagan Railway was constructed from Sicamous to Vernon and four miles south to Okanagan Landing. From 1867 - 1895 many of the settlers were second generation Canadians, who had experienced the hardships of pioneer life in Eastern Canada and knew how to cope in the wilderness.

Some of the early arrivals were Cornelius O'Keefe and Francis Barnard. Cornelius O'Keefe came west from his home in Fallowfield, Ontario, to seek a fortune in the Cariboo gold fields. He brought cattle from Oregon to provide food for the miners and camped at the head of Okanagan Lake to give the cattle a rest and to fatten them on the nutritious bunch grass that grew as high as a field of wheat, and decided to raise cattle in the area of his camp. He pre-empted 160 acres of land eight miles north of the future site of Vernon. By 1907 his land extended almost to the Vernon City boundary when he sold 5,700 acres to Belgian Capitalists.

The O'Keefe Ranch, although in historic form, has survived and is an educational public attraction for those growing up to-day, and a nostalgic experience for those who remember the early days.

Summer of '91 my father, mother and I visited the O'Keefe Ranch. We strolled the grounds and had lunch in the O'Keefe Ranch house. Today, it is a place to visit to relax in and to remember.

Francis Jones Barnard was born in Quebec City on February 18, 1829 and arrived at the Fraser River gold fields in 1859. In 1860 the gold fever in British Columbia brought the first settlers who realized the real potential was in the lush valleys of the Okanagan Valley. Barnard became the earliest rancher in the Vernon area when in 1864 he purchased 6,300 acres of crown land north east of the future town for the purpose of breeding and training fine horses. He named the ranch BX being an abbreviation of Barnard Express. Francis Jones Barnard played an important role in politics. From 1866 - 1870 he was a member of the provincial government and was one of the prime movers in negotiations for the entrance of British Columbia as Canada's sixth province. In 1880 Francis Jones Barnard suffered a stroke. He did not seek re-election and declined appointment to the senate. He was an invalid until his death in 1889. When his father became ill in 1880, his eldest son, Francis Stillman Barnard took over management of his wide business interests. He too was active in politics being elected to the House of Commons for Yale-Cariboo. Although neither lived at the BX Ranch, through it they were important to the development of the North Okanagan. Until the city of Vernon decided to use numbers instead of names the main street of the city of Vernon was called Barnard Avenue. Some people still call it by that name.

Charles A. and Forbes G. Vernon were brothers, born in Ireland at Clontart Castle near Dublin. They came to the Okanagan and for a short time they mined on Cherry Creek in 1863. The Coldstream Ranch as it is named today, was founded in 1864 by Charles A. and Forbes G. Vernon who named it the Vernon Ranch. It was located four miles east and south of Vernon and is an operating ranch today. In 1872 Forbes G. Vernon became the sole owner and in 1891 he sold the ranch to Lord Aberdeen who renamed it the Coldstream, because a very cold stream ran through it. The ranch consisted of 13,000 acres on which were two thousand head of cattle.

In 1884 Charles A. Vernon built a two-story log building with a small store in one corner and stables at the rear of the building. It was located on the north side of what became the 3400 block of 30th Avenue. It was more of a boarding house but Charles A. Vernon named it the Vernon Hotel. As the Vernon Hotel was the first commercial building around which the town grew; the title of founder of the city belongs to Charles A. Vernon.

When a post office opened in 1884 it was given the name of Priest's Valley in honour of Father Paul Durieu of the Oblate Fathers who brought Christianity to the Indians and erected a cabin near Swan Lake Creek a location of the Ukrainian Village (3507 - 27th Avenue). The cabin rotted away and was destroyed totally by fire in 1889. In 1887 the name of the post office and of the community became Vernon, in honour of Pioneer rancher Forbes G. Vernon.

Vernon was incorporated as a "city" in 1892. Forbes G. Vernon, Francis Stillman Barnard and Moses Lumby proposed the spur railway line from the Canadian Pacific Rail Line at Sicamous to Vernon.

The Okanagan Indians lived at the Head of Okanagan Lake. Chief Pelga-mu-lox was the first chief to bring white men into this country and the country of the Okanagans, who were a powerful tribe when the fur traders first came among them. He married an Okanagan woman and the Nicola valley is named in honour of their son N'Kwala afterwards called by the Hudson's Bay factor as NICOLA.

Chief Pelga-mu-lox addressed the white people saying: "You are my white children" I want you to live in my territory. I have a big country, big enough for all of us. Our mountains are green and full of fruit, we have many roots for food. We have grouse and many other birds for the hunt. We have plenty of deer for meat and hides. And salmon to eat. From my water I drink, you drink. From my fruits I eat, you eat. From my game I eat, you eat. You are my white children. Stay with me".

Chief Pelga-mu-lox was a powerful speaker and told other chiefs wonderful stories about the white man. Another chief called his stories false and shot Pelga-mu-lox with two arrows. Chief Pelga-mu-lox was buried on his winter quarters at the Head of Okanagan Lake. His young son N'Kwala was born around 1780. He was the chief of all the Okanagan and Colville Indians.

N'Kwala was noted for his wisdom, prudence and fair dealings by both his own people and the white men. He was a kind humane man. N'Kwala was succeeded by his son, also named N'Kwala. N'Kwala, father and son, were known as

"Great Chief N'Kwala" and were buried in the same grave in the cemetery at the Head of the Lake Reserve.

A fine marble tombstone was erected on the grave which reads:

NO 1 Okanagan - in loving memory of Inquala the Great Chief of the Indian Tribe who was living and ruling in 1838.

The Okanagan Indians realize there can be no return to the life of their ancestors. They have learned to accept the best of the twentieth century while respecting the traditional beliefs of their ancestors. Great Chief N'Kwala would be proud of them.

Development of the Fruit Industry

In 1907 when the Belgian Capitalists purchased land from Cornelius O'Keefe, their acreage circled Swan Lake and the higher range land above the lake and extended southwest from the city of Vernon to Okanagan Lake. They subdivided the land and sold it in small parcels to wealthy Belgians wishing to emigrate to Canada.

In 1910 a Scottish-Canadian Fruit Lands Company composed of Capitalists from Glasgow, Scotland, purchased 120 acres on the east shore of Swan Lake and a 50 acre bearing orchard in the Coldstream area.

In 1911 representatives of several wealthy Paris, France, gentlemen invested in a block of 256 acres purchased from the Land and Agriculture Company.

The early orchardists, including Lord Aberdeen, drew heavily on funds from home to finance their orchard operations. When war broke out in 1914 these funds were no longer available. The young men enlisted. Many never returned. Everyone was short of cash, and short of labour to work their orchards. They were hard hit by the First World War, and by the economic depression that followed. Some gave up their orchards and used the lands for other purposes. Some sold their lands and went on to other careers. Some kept their orchards until the markets for fruit improved. The early settlers developed the expensive irrigation system, without which there would be no fruit industry. There are always ups and downs in the industry. It is affected by weather, disease and problems in marketing.

In 1915 a tent camp was set up on Mission Hill for the recruitment, and training, of men from a large area. By 1916 there were more than seven thousand men under training. The war brought many changes. For the first time woman worked outside the home, in packing houses, in the cannery, and in the dehydrating plants which produced food for the war effort.

William and Katharine at O'keefe Ranch - 1991

Chapter 20

I wrote a brief report of the pioneer years of Vernon, British Columbia to have some knowledge of the foundation set by those early pioneers and of my father's respect and admiration towards them; which explains his desire to live, to love, and to remain in his chosen Apple Valley to his end. He loved the city of Vernon and he loved the whole of the Okanagan Valley; the orchards and the people, the mountains, the forests, the lakes and the rivers. He often took his sons horseback riding through mountain trails. And, in the winter time, he would get a team of horses and a sleigh from the BX Ranch, and he would take his whole family sleigh riding through snow covered woods. There isn't a lake, a river or a mountain he did not visit or camp by with his family from Vancouver, Osoyoos, Trail, through the Cariboo to Dawson City and Hudson Hope.

Vernon was his home.

The foundation of Vernon was important to my father. To him; Barnard Avenue remained Barnard Avenue long after the name was changed to 30th Avenue. And the O'Keefe ranch was, and is a ranch he had often paid visit to.

First they all settled in a mansion surrounded with orchard and garden, with Swan Lake at one end of the property. A year later he owned and operated a new and second hand store on the main street of Vernon, "The B-Y Store".

Spring of 1944, father bought a second hand store from David Rush (where Sweet 16 store is today on Barnard Avenue). Fanny Rush, David's mother was the proprietor.

Father sold his orchard and home on Kamloops Highway and managed his store known as the B-Y Store, until 1965. Father said, and I quote, "I knew nothing about being an orchardist or gardener and I wasn't very successful." He said he had to borrow money to buy stock for the store but he slowly made a success of the venture. He said, and I quote "I liked dealing with the public and met all kinds of people, the majority of whom were honest. I bought antiques and sold them for next to nothing not knowing their true value, but I do not regret that today" he said, when he retired in 1965.

Summer of '44 did not go well for us in our mansion on a hill. Father's injuries kept him from guiding us through the summer with our orchard and garden work we were not too familiar with. With the loss of our three milk cows in the early springtime, there also were no tomatoes and beans for Bulman's Cannery, or a vegetable garden for our own use this year of 1944. If the depression years on our Alberta farm, were difficult years for my father and mother; perhaps I was too young to know. These were the hard times - here in this Paradise Valley.

Summer days in the Okanagan Valley were very hot and dry. The irrigation system the early settlers built was to let water run through the fields during the dry season. Late one evening after working out in the garden all day long in the hot sun, Edward and I went up to the irrigation control system to let some water run into our orchard and garden, through the night. The soil was very dry and the water we released was being soaked right up, it did not run along the flue towards our orchard and garden as it was supposed to do. So; we turned the water control on full and left it on full for the night. In the morning, all the tomatoes and beans for Bulman's Cannery, and all our garden vegetables planted for our own use were at the bottom of our hill in a heap of mud.

Father was stalwart, he did not sit around nursing his injured left leg and knee for very long; when 1944 springtime came, he had a building, he had purchased, in downtown Vernon. A whole block, a concrete building fronting Barnard Avenue, and the back of the building was Coldstream Avenue. And, that spring of 1944, he bought a one ton G.M.C. truck to move us all, his family of nine, into downtown Vernon. How he managed financially, and with a family of nine, is a miracle, a miracle beyond words. Here was a man who as a boy, was growing up during the early Pioneer Years who's father had died while building Canada when this boy (my father) was born. He, my father was never on Relief, or on Welfare, or in a Foster Home. All through his life he was independent and proud. And here, the spring of 1944 we had just surfaced from "Ten Lost Years" the hard times - "The Depression Years". And somehow, my father always found a way. He was a "Survival Genius". Had he written a book, or taught the world "Independent Survival"; today this world would be a better place for all. But then, nobody would have listened. If government officials really cared about our young people; they would stock the school libraries with many books on "The Ten Lost Years" - by Barry Broadfoot for the young to learn about independent survival. They would not be stocking school libraries with the senseless books on "Two Moms" or "Two Dads" which is insane. One must know the past, to build a solid future.

The building my father purchased in downtown Vernon in the spring of 1944; was a 50 year old concrete block building. It was built in 1893 by Alexander Gilmore who owned the building until 1904. The assessed value of the building in 1893 was $2,000.00.

1904 - 1907 the assessed value was $5,000.00 - owned by S. Polson

1907 - 1924 the assessed value was $3,000.00 - owned by S.A. Shatford

1924 - 1926 the assessed value was $3,000.00 - owned by City of Vernon

1926 - 1937 the assessed value was $3,000.00 - owned by D.G. Craig

1937 - 1939 the assessed value was $2,500.00 - owned by City of Vernon

1939 - 1945 the assessed value was $3,800.00 - owned by Fanny Rush

1945 - 1953 the assessed value was $4,000.00 - owned by William Yawney and John Balash

1953 - 1965 the assessed value was $7,440.00 - owned by William and Edward E. Yawney

When my father first purchased the B-Y Block, 115 Barnard Avenue, Vernon, British Columbia, he borrowed private cash money from John Balash for the stock for the store, and he kept John Balash on the title until he was able to pay him back; the loan plus all the interest. After John Balash was paid in full in 1953, the title of the building was put in the names of William and Katharine Yawney. Katharine had it changed to father's first born son, Edward Ernest, and William Yawney, who remained on the title until the B-Y Block was sold February, 1966. The B-Y stood for Balash-Yawney.

Father rented a pasture in downtown Vernon, on 43rd Avenue for our two remaining cows. He sold our horses, Dolly and Dan, the pigs, the ducks, and he kept some chickens. The store had a living area in the back of the B-Y Store facing Coldstream Avenue; a large kitchen, a breakfast nook, dining - living room, full bath and two bedrooms. In the yard along Coldstream Avenue was a grassy area, a big tree and a little creek nearby. My four brothers tied ropes to the tree branches and they swung over the creek from the tree to the other side. There was parking near the entrance door for three to four cars and the truck; a large garden shed where we stored our wood and coal, all our lawn furniture and our camping equipment. We had a little vegetable garden by the shed, and a pen for our chickens. In front of the livingroom window facing Coldstream Avenue was a grape trellis loaded with sweet green grapes. My mother was happy with her home at the back of the store; she liked being near father, to help him in the store and to be the store clerk when father went to an auction or to a private sale.

When my father purchased the B-Y Block building in the spring of 1944, half of the building was rented to a Chinese restauranteur; and that half remained a restaurant into the mid '50's. My mother said that the Chinese man renting half the building showed much gentleness and respect towards them and the children and that he was kind and generous; for that reason they allowed him to stay on for as long as he wished. The other half of the B-Y Block, the Barnard Avenue half, was the B-Y Store. The Coldstream Avenue half was the living area space. The breakfast nook was beside the entrance door to the store; and the store side of that door was a private washroom. A washstand and a crapper. Yes! A crapper. A man by the name of John Crapper invented the flush toilet and during those early days, the flush toilet was referred to as the Crapper, a place to sit, relax and to read the morning paper. There was a long showcase counter that started near the

Barnard Avenue window and front door entrance to near the living area entrance door. The showcase part of the counter had a glass top and shelves for jewellery, necklaces, watches, rings and things people traded in for furniture for their homes. The wall behind the counter were shelves to the ceiling stocked with gloves, socks, underwear, shirts, pants, scarves, hats, caps, and there were skates, boots and shoes. On the opposite side wall of the store were musical instruments of every make, and at the Barnard Avenue front window in front of all the musical instruments, father rented a little space to a jeweller, Mike and Tiny, who did repair, sales and service. The centre of the store was for furniture. New and used furniture to furnish a complete home. There were refrigerators, stoves, washers and dryers, kitchen tables and chairs. Dishes, pots and pans. Diningroom suites, bedroom suites, antique furniture, coat racks, mirrors, lamps and luggage.

Father worked hard to keep his store stocked because he sold for cheap to accommodate people who were low on cash, and because his furniture space in the store was not very large. He went to every auction sale held in Vernon to buy household goods, and he went to every home who advertised a sale of all their household goods, because they were moving on. Near the washroom and the counter in the store, stood a round wood and coal heater with a few chairs beside it, a brass spittoon and a large free standing brass ashtray. My father didn't smoke, but many people who came to the store did. On a support pillar near the heater was an old fashioned wall telephone with the number 833. On the counter near the telephone was an old fashioned gold coloured cash register. I believe that this building, 115 Barnard Avenue that my father purchased the spring of '44; had not been scrubbed since the day it was built in 1893. It was very dirty, and before we moved in, we scrubbed. The floor was wooden planks that had been oiled through the years and they were black. My sister Anne and I scrubbed that floor on our hands and knees with wire scrub brushes until those planks were almost white. We also scrubbed the walls and the ceiling of the store with soap and ammonia then we painted them all white. The ceiling was ten to fourteen feet high, we had to use orchard ladders to reach it. When we got through with the scrubbing and painting, the store again looked brand new. While Anne and I were scrubbing the walls of the store, Mother sprayed the complete living area with D.D.T. She said; just in case there were mites or bugs there. And, she was not wearing a mask while spraying, not realizing the danger of D.D.T. We then scrubbed all the rooms with soap and ammonia, then painted them all white too. With all the walls and ceilings painted white, the whole place looked bigger, brighter, white and clean. The floors in the two bedrooms, the bathroom, kitchen and breakfast nook were covered with old linoleum, and the living - dining room floors were some sort of a carpet. Father ripped up all the old floor and coverings and he put down all new material. And then we all moved in. In the store, he painted the floor with fresh clean oil and the store was opened for business.

The Farmers Market Store Edward and I walked to from our orchard farm on Kamloops Highway to sell our farm fresh eggs, cheese, butter and cream; and to buy corn flakes and ice cream, was a few doors down from the B-Y Block on the same side of Barnard Avenue at 34th Street. It was the “Bagnall Block” of a wooden structure. 1933 - 1958 the business was “The Egg and Poultry Exchange”. Everyone in Vernon referred to this business as The Farmers Market Store until

1958 when Vernon was putting on a new face; the business changed, though the Bagnall Block remained. Street names were changed to numbers and my fathers B-Y Block; from 115 Barnard Avenue became 3308 - 30th Avenue and his store telephone number became 2833.

B-Y Store - World War II era. - 1945

Coldstream Avenue entrance

Barnard Avenue in Vernon, British Columbia

Coldstream Avenue entrance to B-Y Block and Fargo truck - 1949

Back of the B-Y Store Katharine's mother sitting

Grandmothers visit father's "Apple Valley"

Chapter 21

Winter of 1944, Anne and I worked in the Okanagan Café on the weekends. Anne was serving the public, and I was in the kitchen working as salad and sandwich girl. While the B-Y Block was being renovated, we stayed downtown in Vernon on Friday and Saturday nights with an old lady. Her name was Mrs. Shuldes and she lived alone in a suite in a big house, the Vernon Lodge. A house where today stands the popular Vernon Lodge Hotel. She had no family in Vernon and she enjoyed having us stay with her. She loved reading tea cups, and she showed us how to read each others tea cups. She often invited ladies to her place for an afternoon tea, and she read all their tea cups - no charge. Anne and I worked the late shift in the Okanagan Café on the weekends and after our work was done, we walked to her place, the Vernon Lodge about midnight. That house she lived in had a big front verandah with a swing to sit on and to watch the people passing by. A magazine rack stood by the front door to collect the morning paper, and a table was there for the bottled milk to be placed on, milk that was delivered during the night by a milkman and his one-horse drawn milk wagon.

One night after we worked, then walked that distance to that Lodge, we were not aware that we had been followed by a couple of young Army Cadets. They sat on the swing on the verandah all night, and in the morning they read the morning paper and they drank up all the milk that was delivered to the Lodge. Perhaps there were more than just two Army Cadets who spent the night on that verandah, but that is what we were told, that there were two.

When Mrs. Shuldes told my father about this, he immediately stopped us from working weekends and from staying downtown. Anne and I got our jobs back in the Okanagan Café after the family moved into town and our home was at the back of the B-Y Store.

Father was our mentor; because of the Army camp in Vernon with thousands of cadets there year round; he kept an eagle eye on us. He gave us a key to enter the store, after work and not to our entrance door on Coldstream Avenue. There was a big window in the master bedroom which looked out into the store, and out onto Barnard Avenue. Father always knew the hours we worked and when we were expected to be at home. And from his window he could see if we were being followed. The Okanagan Café was two blocks up and on the same side of Barnard Avenue as the B-Y Store. Anne continued working in the Okanagan Café late into the summer, and after school was out for the summer, Edward, Henry and I went out to the farms to pick tomatoes and beans. There were many farms along the Kamloops Highway from Vernon, through to the O'Keefe Ranch who hired us. They picked us up in town, delivered us out to the farm, to the job, then brought us home again in the evening.

When the B-Y Store opened its doors for business; this was father's own project, an adult responsibility. To serve the public was no place for children to take part in, or to loiter in. Customers who couldn't control their children in his B-Y Store were asked to leave. He said a store is a business not a play area. Before long all the people of Vernon were aware of, and respected father's business rules, and when they brought their children to the store, the children were kept in control.

Across Barnard Avenue, from the B-Y Block where the Woolworth's Department store is today, was an empty space. It was like a park, a grassy area with big trees. When the Indians came, they tied their horses to the trees there and they left their wagons in the shade of some trees, then they would help each other across Barnard Avenue to the B-Y Store. Most of them were older people; some were scarred and crippled. They were regular customers of the B-Y Store. Father was kind to them, and he talked to them about many worldly things while they sat on the chairs near the wood and coal heater. They enjoyed conversation with my father and he with them; and they knew they were welcome. Father talked to them about education; to educate their children whatever the sacrifice. The Marchand family took my father's advice and they sent their son Leonard Stephen Marchand to University who became a Senator in Ottawa in 1984 and still is today. If it was my father who convinced this young lad to go on learning, or if it was Leonard Marchand's own choice, I believe there were no regrets. My father was proud of him.

Father's sons, Edward and Henry, were students at the Vernon Polson Park High School when Leonard Marchand was a student there the only Indian. Together they studied and they played lacrosse, then again during the mid 50's they were at the University of British Columbia continuing their education. From the university they each went their separate ways. I was extremely pleased and grateful to have recently talked to Honourable Leonard Marchand P.C., on the telephone and to have received from him from Ottawa, a copy of his biography.

BIOGRAPHY

THE HONOURABLE LEONARD STEPHEN MARCHAND,P.C., B.S.A., M.S.F., P.Ag.

Senator Marchand, an Okanagan Indian, was born in Vernon, British Columbia on November 16th, 1933. He is the son of Agnes and Joseph Marchand. On July 16th, 1960, he married Donna Isabelle (Parr) of North Bay, Ontario; they have two children, Lori Anne and Leonard Stephen Jr.

Mr. Marchand spent his early years on the Okanagan Indian Reserve where he completed grades one through eight at the old, one-room, school at Six Mile Creek. He attended Kamloops Indian Residential School for grade nine and Vernon High School for grades ten through thirteen. He was the first Indian to graduate from Vernon High School when he completed grade thirteen in 1955. He attended the University of British Columbia and received a Bachelor of Science Degree in Agriculture (BSA) in 1959. He continued his studies in range management at the University of Idaho and in 1964 completed a Masters of Science Degree in Forestry (MSF).

In the late fifties and early sixties, Mr. Marchand was an active member of the North American Indian Brotherhood. He worked on various issues such as obtaining the federal vote, self government and improved education for Indians. From 1960-65, he was a research scientist at the Agriculture Research Station in Kamloops, British Columbia where he was author and co-author of several research papers.

In 1965, he became the first Status Indian to work as a Special Assistant to a Cabinet Minister, The Hon. J.R. Nicholson. He was also a Special Assistant to The Hon. Arthur Laing from 1966-68. Mr. Nicholson and Mr. Laing were both Ministers of Indian Affairs.

In the general election of 1968, Mr. Marchand was elected Member of Parliament for Kamloops-Cariboo, British Columbia. He was the first Status Indian to be elected to the House of Commons and was re-elected in both 1972 and 1974. He was appointed Parliamentary Secretary to the Minister of Indian Affairs and Northern Development, The Hon. Jean Chretien, from 1972-73, and the Minister of Environment, The Hon. Jeanne Sauvé, from 1974-76. In 1976, Mr. Marchand was sworn into the Queen's Privy Council when he was appointed Minister of State for Small Business. He was also Minister of State for the Environment from 1977-79.

During 1979-84, Mr. Marchand was the administrator for Nicola Valley Indian Bands in Merritt, British Columbia and Director of the Western Indian Agricultural Corporation, an agency designed to encourage advanced agricultural production methods among Native Canadians. He was also a private consultant on government, agricultural and aboriginal concerns. As Chairman of the Round Lake Treatment Center Building Committee, he helped establish Canada's first Native drug and alcohol treatment center, near Vernon, British Columbia.

On June 29th, 1984, The Rt. Hon. Pierre Elliott Trudeau appointed Mr. Marchand to the Senate of Canada to represent the province of British Columbia, Kamloops-Cariboo designation. Senator Marchand has been active in many Senate Committees especially those dealing with social and aboriginal affairs, agriculture and forestry, and internal economy. On February 20th, 1990, he was the first Chairman elected to the newly formed Standing Senate Committee on Aboriginal Peoples.

Senator Marchand was made Honorary Chief of the Okanagans in 1984. He is also a member of the Liberal Party of Canada,
Board Member of Peace Hills Trust Company, Director Emeritus of the Canadian Council for Native Business, and Honourary National Patron of the Canadian Native Arts Foundation.

He is an avid sports fan and enjoys reading, horseback riding, and playing pool.

Edward always had many native friends on the North Okanagan Indian Reservation. People he got to know at the B-Y Store from the years the store first opened; and he often went to visit the families on the Reservation, where he spent many hours horseback riding with the Indians along the lake trails. Henry, Patrick, and Marshall as they were growing up, also spent many hours riding Indian Ponies on the Reservation, and they all said, and I quote, "The Indians were all good people". They were farmers, cattle ranchers, they made hay, and some of the young native men worked on the O'Keefe Ranch year round. The names of some of the natives who came to the B-Y Store, and who were visited in their homes on the Reservation are: Marchand, Edward, Brewer, Bonneau, Lazard, Gregory, Bissette, Williams, Antoine, Cameron, Lewis, and there were more.

After one year at the University of British Columbia, Edward attended a Normal School then began teaching in West Vancouver in 1957. Taking a break from teaching during the summer months, Edward worked on the Perry cattle ranch on the Trans-Canada Highway eight miles east of Cache Creek driving towards Kamloops. And, there were many summer times he worked as a cook on the C.P.R., traveling to Winnipeg and to Toronto. After a few years of teaching in West Vancouver, then in the Scow Street School in Burnaby, he transferred himself to Kamloops then on to Quesnel, British Columbia "The Cariboo" where he bought a cattle ranch. He then returned to the University of British Columbia for his Bachelor of Education degree, after which he was a School Principal in Quesnel until he retired in 1991 and returned to Vernon when my father's strength and eyesight were failing him. When Edward returned to Vernon, he resided on a beautiful 6.3 acre hobby farm on Scott Road, in the city of

Vernon; acres my father had, and had purchased when he was still operating the B-Y Store. The acres were fenced and cross fenced with a natural spring from an artesian well drifting through the property to the pasture below. There was an older two bedroom stucco home there, barns, chicken house and pen, and garden sheds. There were Italian prune plums, crab apples and lilac trees by the house and tall Russian poplar trees by the barns. It was a beautiful spot; from these acres we could see the Kinsman's Beach and Lake Okanagan, the Vernon Airport and the mountains beyond.

Henry in his third year of medicine at the University of British Columbia, left the University to marry his lady Joan. He then was kept very busy travelling across Canada working with doctors as a pharmaceutical representative. After his family were all moved to Grimsby, Ontario and he traveled all of Eastern Canada visiting all the doctors and most of whom were in Toronto and in Montreal, having to be away from his family for several days or weeks at a time, he left his position as pharmaceutical representative and he returned with his family to British Columbia, to Kamloops, where he bought a new home and went into business for himself. He bought a fleet of dump trucks and was hauling sand and gravel to road and building sites. For his family, his wife Joan, his five sons and three daughters, alone, he built a beautiful summer cottage on the banks of the Shuswap Lake. Henry's desire was to always be with his family; he did not mind what kind of work he had to do, as long as his family were near. During the years he was a student at the University of British Columbia, he worked at J.R. Murray's Plywood Mill in Marpole; and he worked at Clarke Simpkins Motors, as a car jockey.

Father was alone with his B-Y Store, a one man case load to provide for us all and alone he paid all the costs and tuition fees at the University for both his sons Edward and Henry. The jobs, or work Edward and Henry did during the summer months were of their own choosing. Our home and way of life in Alberta, and on our orchard farm on Kamloops Highway, were a family affair; we all took part in the work that had to be done. The B-Y Store was a different way of life. As the business grew, the one ton G.M.C. truck became much too small for the stock father had to bring in to the store and for the trips he made once or twice a month over the Hope-Princeton Highway to Vancouver to buy stock. He bought furniture, stoves, fridges, washers and dryers at Love's Auctioneers and Appraisers Ltd., 1635 West Broadway and he bought antiques and quality furniture at Maynard and Sons of Vancouver Ltd., 1233 West Georgia. When his truck was not full, he paid visit to Ross Auctioneers and Appraisers Ltd., 3890 East Hastings, Burnaby, and at Tyldesley's Ltd., 4515 Main. The boots, shoes, trunks and luggage he got in Gastown from the wholesalers. When father started making frequent trips to Vancouver; he traded his one ton G.M.C. truck for a 3 ton Chevrolet. The boxpart of that truck was very high up off the ground and it did not have a hydrodynamic lift. Loading the furniture and appliances onto the truck and unloading them into the store was all done with arm, back and leg muscular power. Father had a 600 lb. dolly to cart the heavy appliances from the truck into the store, and from the store to the truck when he made deliveries. A dolly is a flat deck with two little wheels at one end of the deck and two four foot long handles joined at the top going straight up from the two little wheels. The flat end of the deck without wheels was pushed under a fridge or stove and that appliance was then strapped on to the handles, then tipped slightly onto the little wheels and pushed into the store. This same procedure was used to cart heavy appliances out to the truck when they were to be delivered to his customers. Father did much of the heavy lifting all by himself. He did not ask his sons to help with the lifting; he did not want them to strain their young bodies, lifting heavy furniture. The lifting part, on and off the truck, was always done by man power, and father never had to advertise for help to unload the B-Y truck when he returned from his many trips to Vancouver. Young men around town searching for a way to make a dollar, always knew when the B-Y truck was out of town, and they watched and they waited. When the truck arrived, they were right there to help father unload, and to get paid their beer money, then off to the Vernon Hotel they would go to satisfy their thirsts. Father was pleased to have help waiting when he arrived with his load after his long all-day drive over the Hope-Princeton Highway.

a long hard trail a-winding
into the land of his dreams

Many times father arrived with his load late in the night, and alone he worked, by the light of the silvery moon unloading his B-Y truck while we all slept in our comfortable beds. The days he arrived in the afternoons, and there to see his men waiting to help, worn and tired he smiled, as his heavy load was lifting. The B-Y truck was quickly unloaded, the men helping were paid their beer money, and everyone was happy.

Today; I realize, that during the years of the B-Y Store, father's Katharine, his sons and his daughters did not see, and were not aware of the hard and heavy work father alone was doing, to keep his B-Y Store proper and his family well supplied.

Chapter 22

The Okanagan Valley was a beautiful Valley; from the North Okanagan Lake and the O'Keefe Ranch, the Swan Lake and Vernon, along the Kalamalka Lake of many colours, to Oyama, Kelowna, Westbank, Okanagan Mission, Peachland, Summerland, Naramata, Penticton, to the South Okanagan Lake, to Okanagan Falls and Skaha Lake, Oliver, Osoyoos, to Keremeos, Hedley, the Similkameen River to Princeton.
This was my father's Apple Valley.

Every springtime, father took his family out driving in his Plymouth car along the Okanagan Valley Highway to see and to enjoy the aroma and the beauty of all the different blossoms, on all the different fruit trees. There were Bing cherries and Royal Anns, peaches, apricots, pears, Italian prune plums, and a variety of apples. As the seasons changed, father took his family on Sunday drives through the beautiful Okanagan Valley to pick, and to buy the ripe fruit in season. We could pick and eat the fruit right off the trees. This was before all the fruit trees were being sprayed with chemicals and D.D.T. Most of the chemical spraying started in Canada in 1949. D.D.T. was banned in 1972, and in the '70's and the '80's all the cherry trees were worm infested and all the apple trees were chemically soaked through and through.

One late Sunday afternoon as we were driving home towards Vernon along that beautiful Okanagan Valley Highway from Keremeos with our baskets full of Bing cherries; those cherries were big and beet red, and they were sweet and juicy, with our car windows open wide, we were enjoying the Okanagan evening breeze when suddenly a "Whack-o'; my mother got smacked in the eye and blood red was streaming down her face. She grabbed her face with both her hands and she yelled - "I have lost an eye", "something flew in through my window from the outside and it hit me in my eye". Father quickly stopped the car and he gently took both her hands off her face; and he said, "Your beautiful blue eyes are both there, Katie. You have Bing cherry juice all over your face." A car driving along the highway in the opposite direction, tossed a Bing cherry out their window, the cherry flew into our car through the open window and it smacked mother square in the eye. There was no real damage done, just her sudden fright of losing an eye.

The Okanagan Lake was thick full of Kokanee fish during the '40's. One evening after father closed the B-Y Store, he took a metal tub and our canning boiler and he went to the banks of the Okanagan Lake where the Koknee gathered in thick schools. Within a half hour he returned with the tub and the boiler full of delicious fresh red flesh Koknee fish. The Okanagan Lake was truly full of Koknee fish during the '40's: Today there are none. From all the orchard chemical spraying, the chemical run off from the orchards into the Okanagan Lake through the years, must have destroyed all that delicious Koknee fish.

Today this is happening to all our fish and shellfish; not only in the lakes, but in the oceans too. There is too much pollution, and too many chemicals dumped into our lakes and oceans. Today people are a walking chemistry, and they are dying off at a younger age, from the old age our hard working pioneer settlers did. The pioneers worked hard and under very poor conditions, but their food and their water was pure and clean and free of chemicals. All our dairy cows are being injected with a chemical substance today to make the cows produce more milk and to be fed less hay. The chickens, pork and beef we consume are also chemically fed to make them grow faster, bigger and more tender. What has education been doing to this world? Is it all for the sake of the money, above all else? The knowledge of science and psychology has often been used, and is often being used, for destruction, and for the money, not for the betterment of mankind.

Today there is no kindness, compassion and trust amongst people. People today do not care for, nor do they help each other, as they used to. Americans, USA and Canada, are interfering with other countries, telling them how to run their countries, how to live, how to treat their children and their women, when those rules should be left alone, for each country to do as they see fit. That is their house, not ours.

There were no locks on our doors in our homes during the '20's, '30's and '40's and there was no need for locks or for locking our homes. Today homes are locked and double locked and are being broken into. Have our past and present Prime Ministers allowed too many different breeds of people to enter our Canada? I believe so. Since the 1970's, the people who were allowed to enter Canada, to this day, are trying to make our world into their own kind, while our taxes are paying for the best health care and education for them. A lot of the money they earn here in our Canada is sent back to their own countries, and that is a well-known fact. All the people who have entered our peaceful Canada during the '70's, '80's and '90's must think we are a bunch of numb-skulls. If we are not able to make our Prime Minister wake up and to see what he is allowing to happen, we will be over-powered by these people and there will be nothing, nothing at all that we will be able to do. The early Americans, when they first discovered America, fought our peaceful Native Indians for this, their native land. Today, Canadian soil is being taken over by people from other countries who are changing the appearance, and the quality of life here in our Canada. During the pioneer years, Canada was a French, Scottish and English Canada. All other nationalitics who entered, respected and abided by their rules, and demanded nothing. They were happy to be Canadians, to have a piece of Canadian soil, to live on, to build on, and to care for with all their strengths. The foundation they built during those early pioneer years, Our Canadian House Foundation, should remain till the end of time; and not be allowed to change to suit every different breed who enter. The French Canadians are desperately trying to save, and to keep their precious piece of Canada - Quebec - as it was, before they are forced to feel as though they are in a foreign land.

The Beijing China students, who caused a commotion in Tiananmen Square in China in 1989 are troublemakers, and should not be allowed to protest and to demonstrate at our Universities and on our city streets. Are they Canadians? Or are they Chinese? And why are they doing their protesting here in Canada? I believe the Beijing Chinese students were troublemakers in China; and are troublemakers here in Canada; and they should be sent back, to be dealt with their arrogance, in China. That is where their trouble began, and that is where it should end. Canada should not accept and shelter students from other countries. The Beijing China students were a group of privileged citizens in China, who are ungrateful to their people and to their country.

The 1895 - 1914 pioneers who came to Canada from China; were, and are today a beautiful, peaceful hardworking people; not at all like the recent arrivals. Canadian people known to me, acquaintances, who have recently travelled through all of China; including Beijing have said, and I quote; that China is a very well run and organized country, and is very beautiful.

September 1992, my mother, brother Henry, Rita Hughes a retired medical professional; and their friends travelled all of China, to the Great Wall and Beijing. My sister Anne with her teacher friend Agnes Cerny and a group of Canadian women travelled most of Communist China in 1995. They said that all the hotels in China are very nice and clean, and in well kept beautiful architecture, that their food was always good and their service was always excellent. They said there were always many Army personnel on all their city streets keeping strict orders that all the people they met and talked to had nothing but praise for their country China, and for the way it is run. And they said that Communist China does not need any coaching from Canada and from the United States of America of how to run their country.

Of their travels through India, they said in all of India, there are the very wealthy, and there are the very poor. Two opposite extremes. It is not the poor who have come to Canada and are getting themselves into Canadian politics. Think about this Canada . Think about this very seriously.

Indira Gandhi (1917 - 1984) was the first woman prime minister of India. She held office from 1966 to 1977 and from 1980 until her death in 1984. She was assassinated by two of her security guards who were members of India's Sikh religious group.

In the early 1980's Indira Gandhi had to confront a variety of violent conflicts among India's religious and ethnic elements. In the Punjab, a faction of Sikhs aiming at secession from India murdered many government officials, Hindus and members of the Nirankari Sikh sect, whom they considered heretics. Indira Gandhi's secularist outlook, her identification with the poor, and her image as a strong national leader had made her the only politician of all-India stature acceptable across regional, religious, caste, and class lines. She strove to combat India's poverty without changing thc structure of society. To assure the poor the benefits they were entitled to as equal citizens under the law.

Indira Gandhi had two sons, Rajiv and Sanjay. Sanjay Gandhi became his mother's chief political adviser and gained much power in Indian politics. Since the early 1970's Sanjay Gandhi, a forceful and controversial politician with strong support among urban youth, had been Indira Gandhi's political lieutenant. He was killed in an airplane crash in 1980??? Rajiv Gandhi then became a key aide to his mother. Indira Gandhi persuaded her older son Rajiv, to enter politics as her main deputy. Rajiv was a good listener and a fast learner. His quiet calm manner inspired confidence. Rajiv had only four years of political apprenticeship before, on October 31,1984, Indira Gandhi was assassinated. Rajiv was chosen prime minister by cabinet upon her death, in the next days Hindu mobs killed thousands of Sikhs. Rajiv Gandhi then called general election for December 1984 and, led the Congress Party to its greatest victory ever. While campaigning for the ensuing elections, Rajiv Gandhi was killed by an assassin's bomb on May 21, 1991.

Native Canadians, Pioneer Canadians, think and think again very seriously about our Canada, our land, our Royal Canadian Mounted Police, our politics, and who should be allowed to enter our country, before our parliamentary elections. A growth, a drastic change, is a cancer and must be stopped.

During the '40's, '50's and '60's we Canadians, were able to travel anywhere and everywhere in Canada, and we were able to pitch a tent at any lakeshore, or on any remote riverbank and camp for several weeks at a time, or for a weekend, and enjoy the outdoors, the wild animals, the birds, and Mother Nature. Today, the '90's, we do not go camping and fishing in the wilderness to enjoy the outdoors, to dance and to sing around a campfire, to roast wieners and marshmallows over an open fire, because there are dangerous and strange people who are wandering around. We do not fear the wild animals in the wilderness, we are terrified of the people. The people who were allowed to come to Canada and to live amongst us.

Our second, third and fourth generation Canadian children are missing our outdoor camping life as it was; as we were able to. Today, our country Canada is shameful and cruel to our children of our pioneers, and our Prime Minister Jean Chretien is saying "Canada is the best country in the world". Not today it isn't; not since he has allowed wealthy wicked trash to enter, and who today are beginning to rule. How much wool does he have up his sleeves to pull over those eyes who cannot see.

After 87 years of my father's building and trusting, today his children, grandchildren and great-grandchildren are not safe. Not safe in their homes, not safe in their schools, not safe on the streets, and not safe to go camping, to enjoy the world he was building for them.

Chapter 23

Our home in downtown Vernon, brought us to meeting and to knowing many Okanagan Valley people. Most of the people who came to the B-Y Store, came to see, and to meet the "Hill Billys" who ventured out through the Rocky Mountains in a boxcar.

Many of the people who came to the B-Y Store were curious; they wanted to meet this man, my father, and to talk with him. My father was a broad-minded man; he knew much about life, about living, and about our world, and he listened to people, he did not force his beliefs onto anyone. His personality, and his knowledge about life, the physical hard work, and the intellectual, brought him many customers. He was friendly with all people, though true to his family alone. He never chose friends over family to be with. He took his family on picnics, on camping trips, or just driving the Okanagan Valley highways.

My father was the brave, proud, adventurous and strong, the resourceful man all his customers, and the people he knew all looked up to with much respect and admiration. Many people from Alberta were being curious, and began coming out over the mountains to see how my father was doing with his family of nine. My father opened the way for the many people who came, and every year they came, they bought Okanagan Valley fruit before they returned to their homes in the cities, in the towns and on their farms of Alberta. A curious young man from the Lornedale School area, a farmer's son from a family of ten, arrived in Vernon in 1948 to pick cherries, he said, and there in Vernon he met his Julia. A year later, he turned to my father for financial help. His desire was, to marry his Julia and to take her with him, to his home back in Alberta to meet all his family. Today, he is a tired and crippled old man from working in mines and in saw mills and is most grateful to my father who's financial help made his dreams come true, today with his Julia, they are growing old together in their home in the Garibaldi Highlands at Squamish. He called to say, that through the years he had hoped that my father was aware of the sincere gratitude he always felt towards him for the loan he allowed him for his beginning, with a handshake and a wink of an eye. The loan was soon paid in full, and my father was pleased he was able to make a young man happy. The year 1952, in September my father sold his Alberta farm to this young man's father's brother, a Bill Petruk from Edmonton whom this young man was named after.

The Bill Petruk my father sold his farm to, is the man who demolished our log house, all the log barns, the garage, pump house, sheds and corrals and knocked down all the trees that father left standing, which made a windbreak for the grain; and made our acres beautiful and picturesque. For me, this was sad to see in '93. I was first born on these beautiful acres, which today is no more, but an open field of grain with no trees, no birds, and no buildings my father worked very hard to build. In todays terms, this is progress.

I joined a small orchestra group in Vernon after we were all comfortably well settled in our home at the back of the B-Y Store. Anne, Edward and Henry soon followed. Anne played her guitar, then switched to playing a mandolin. Edward and I played our violins and Henry played a bass

French horn which suited him just fine. His voice was always a loud low bellow tone when he talked, and with his large chest and strong pair of lungs, he brought the oom-pah-pah's out of that French horn right on. We practiced every week and we toured the Okanagan Valley playing at concerts. When we travelled to Penticton, we all stayed overnight at the "Three Gables Hotel". That was exciting for us, our first time ever staying in a hotel, window shopping, having a picnic lunch in the park before being loaded on to a bus for our drive back to Vernon. We also travelled to Vancouver to compete with other orchestra groups. We travelled in a bus, along the Fraser Canyon, and Oh! was that ever a scary journey. The Fraser Canyon was a trail built for horse and wagon with one lane, old wooden bridges stretching from mountain to mountain over canyons. When the bus met another vehicle going in the opposite direction, one of the vehicles had to go backwards until they could find a wider space along that trail for the two vehicles to pass. During the '40's most people travelled by train through the mountains; they did not travel the mountain trails. Our music teacher chose to take his orchestra group, the players and all their instruments in a bus. He must have been the brave one, or he just enjoyed seeing our frightened faces.

October 1944, my sister Anne left the Polson Park High School, her grade 11, and our family home. She was off to the big city, to Vancouver, to find work and a new way of life. Two other Vernon girls went to Vancouver with Anne searching and exploring; she was not alone. Anne got a job at the Woodward's downtown Vancouver store at 101 West Hastings, in the grocery department. As a growing little girl, and right up to the day she left our family home, she always talked, and hoped, that she would become a school teacher. The excitement of going to the big city with friends made her leave home before she could complete her high school grades. She was a beautiful girl with dark hair and blue-grey eyes, a slim waistline and a figure of a model. She could have been a model, though she kept her job at Woodward's for many years. She was a responsible and reliable employee and Woodward's paid her well. Mother was not happy that Anne left our family home. She said she needed Anne's help in the home, and with the children, and she blamed the two girls, Anne's friends for taking her with them. Father would have preferred that Anne stay and complete her high school grades before leaving home, though he accepted her choice. He believed that a person, a son or a daughter, at the age of 17 years, has the right to choose a way of life. His trips to Vancouver became more frequent; he was keeping a close check on his oldest daughter's new way of life, of the work she was doing, and the home and the people she was boarding with. At any time of day; unannounced, he would walk into the Woodward's grocery department to see if his daughter was happy with the work she was doing, and to see what she was doing. I remember a weekend, a weekend trip father took all of us in his Plymouth car to be with Anne for a day.

The Army Camp on Mission Hill in Vernon was always a going concern with Army personnel. Some cadets were always leaving camp, while more cadets were always coming in. The streets of Vernon were full of khaki clothed young men marching along. Their Army training was all done up on the Mission Hill, though when the cadets came into town, they marched along in perfect step. They all were a group of fine young men; never did they cause any trouble to merchants, their customers or to the old folks and children anywhere along the city streets. There were many winter nights during 1945 – 1949 when two and three girls and I were chosen to play our violins for the Officers in the Officers Mess Hall on Mission Hill, with Mrs. Nickols our piano accompanist. Mrs. Nickols was a piano teacher, and an excellent accompanist. We played solos, duets, and quartets for several hours per night, once or twice a week, then we would walk back down Mission Hill into town. I will never forget one winter's night when we walked down that hill with the city lights sparkling brightly through the huge silver-white snow flakes softly falling to the ground. The night was calm and quiet, and the air was fresh and clean. It was so peaceful and beautiful walking along in the falling snow. We walked past the old little Vernon Jubilee Hospital, then past the Polson Park and our high school and on into town and to our homes. I do not remember if we got paid for the dinner music we supplied the Officers with; I believe we did not. Perhaps our manager who arranged the dinner entertainment for the Officers did. We enjoyed playing and being stage performers.

Spring of 1945 during our school Easter exam period, my mother was taken to the Vernon Jubilee Hospital for surgery. I believe it was a hysterectomy, though she never did tell me. Right then and there I was asked to leave school and to take over the responsibility of the home and children; which I did. After her surgery, mother was brought right back home for her convalescence period, and it lasted until well after our school Easter holidays. I cooked, sewed, cleaned house, did laundry every day, took a food tray to mother's bed, morning, noon and night. I removed her bedpans, sterilized them and gave her sponge baths. And I looked after the children; Joyce was three years old, Marshall was five and Patrick turned seven on the 17th of March. Since I was very young, I talked of and hoped that I would become a nurse after my high school education; and here in my father's home, I was getting my pre years nurse's training at a young age of fifteen. I did not mind the work, I seemed to know how, and what to do, and I was relied on. When my mother was feeling stronger and I was released of my duties to return to my classes in school, I went to school that one sad and difficult morning and after all the students were in their classrooms, I went to my locker, took all my school books out, and was near the front door when Mr. Seaton, my homeroom teacher, caught up to me, and he begged me not to quit school. I told him that I could not stay. I was repeating my grade nine that year and because I missed all my Easter exams, I could not accept, or face, failing grade nine again. I could not, and I went home; where my help was required. He said he would do anything to help me catch up to my class if only I would stay. And those words kept coming back to my mind over the years, and because of those words; I will never forget my homeroom teacher, Mr. Seaton.

Every year a "Vernon Days" was celebrated with a stampede. All the stores, shops, banks, etc. were decorated with log slabs covered with bark on the walls to make them look pioneer. There was a parade and a "Vernon Days" queen was chosen, and every year from Vernon, a beauty queen was chosen to attend the P.N.E., The Pacific National Exhibition in Vancouver. August 1945, my mother got a phone call from the P.N.E. to enter me in a beauty contest with all the Vernon beauties. And she was told that the winner will travel, all expenses paid, with her mother to the C.N.E., The Canadian National Exhibition, in Toronto. Well! Patsy Laidman got chosen as first; then there was Doris-Kay Graves, and there was me. I do not remember who came second and who was third. We each received several hundred dollars worth of gift certificates to buy clothes with and I will never forget a very expensive pink angora wool sweater I bought. It was very pretty though everywhere I went I left a trace of pink angora wool; it clung to everything, everybody, everywhere. No! my mother and I did not go to the Pacific National Exhibition. And, that is as far as Patsy Laidman got to, no farther. My mother and I went to all the exclusive shops in Vernon and bought all new clothes for me, and come September, with my violin firmly under my arm, I was off to Vancouver to study music, to continue my music career. When my father was going to Vancouver for a load of furniture for the B-Y Store. I went along with him, and there in Vancouver I moved into a rooming house together with my sister Anne. We had one room with a hot plate and a sink. I took private violin lessons from a violin teacher in his studio for a short period of time. His name was Harry Hoshowsky an extremely good looking man who sent me on to join the Vancouver Philharmonic Symphony Orchestra under the direction of Douglas Stewart, who's music studio was under the clock in the Vancouver Block. And there, with the Vancouver Philharmonic Symphony Orchestra I had the privilege of meeting Glenn Gould pianist, Gerald Jarvis violinist plus many more excellent musicians.

From that day in September my life had changed; from family and hard work to music. Father came to Vancouver often to visit his two daughters, and whenever I knew that he would be in Vancouver during a symphony orchestra concert, I always had complimentary tickets for him and for Anne to be there at the concert. At one time when father was in Vancouver, he took us to see a movie of our choice. I wanted to see "Portrait of Jenny" with Joseph Cotton, Jennifer Jones and Ethel Barrymore. This was a romance movie and I forever will remember it because my sister and I went to see that beautiful movie with our father.

At one time when Father arrived it was warm and sunny outdoors. Anne and I packed a lunch and took father to Spanish Banks, a beach, where we sat on a log and watched an ocean liner slowly moving towards the Lion's Gate bridge. We had no idea that on that ocean liner Anne's "Never Ending Love" Paul Chauvel was coming in. He came and he went many times for several years before Anne met him; he was working as a bartender on the ocean liners coming into port from Australia.

April 1946, for my 16th birthday, by special delivery courier service, I received a beautiful and very expensive violin from my father from the B-Y Store. He furnished a complete house for the people, to get that violin for me. I was so thrilled, and so excited, I could not believe that this beautiful violin was really mine. I phoned my father, and I said that I will work and I will practice and I will become a violinist, a concert master and with those spoken words, I played and I practiced with all my energy, emotion and tears. When I played, the music did not come out of the violin alone, but from my heart, my soul and from me.

I worked at the "National System of Baking" in Vancouver to pay for my music lessons and for my share of rent and food. Anne and I with a girl from Vernon whose name was Rose Yakimovich, who was a ballet dancer, were renting a little house on Marine Drive. This little house we rented was surrounded in bush, a little bit scary, though much better for my constant practicing, which could have been disturbing to people had we continued living in the rooming house, or in an apartment. And for the three of us this set up was perfect; I played, Rose danced, and Anne cooked. I had to leave this house at 4:00 a.m. every morning to be in the bakery, on the job making scones and donuts to be ready for delivery to restaurants at 6:00 a.m. This is where I really had my hands in the dough. There was a path from our little house through the bush to the bus stop. One early morning, at 4:00 a.m. as I was creeping through the bush along that path in the dark of night and in thick fog, I heard loud thump, thump, thumps and branches crackling coming towards me, and my heart stopped. I made an about turn and ran back to that little house as fast as my legs could carry me. Frightened and puffing, I woke Anne and told her what I had heard in the bush. She quickly got ready and together we walked the long way around that bush to the bus stop. That morning, Anne had a long wait before she was able to get into the Woodward's store. After that frightful experience, I often wondered if that creature that was thumping along that pathway towards me in the dark and in the fog was a Sasquatch.

Douglas Stewart our symphony conductor did not approve of me working in a bakery. He said that I should have my hands insured and any work I chose to do, should not be dangerous to my hands. Before I was able to insure my hands and to find another job, I was rushed to the St. Paul's Hospital from the bakery with an acute appendicitis for immediate surgery. When I was waking up from anesthesia, father was there beside me, and when I was able to travel, he took me back to Vernon to recover. In Vernon; I entered the "McEwan-Wilkie Business College" and completed a course in bookkeeping, typing, payrolls, business machines, calculators, adding machines, filing, etc., all office procedure. I asked my mother to come to the college and to take a bookkeeping course, but she did not. While I was in college, I entered the Okanagan Valley Music Festival and won the trophy silver cup for best violinist in 1949.

In the late autumn of 1949, and after my college course

was completed, father took Edward and me, the two who rode with him in a boxcar, back to Alberta to see our birth place and to travel with him along the Big Bend Highway the same route he travelled with the family who were waiting for him on the Alberta prairie without a home while Edward and I were alone in that mansion on a hill in an apple orchard, eating cornflakes, strawberries and cream.

Vernon Days beauty contest

Beauty queens on stampede grounds

Vernon Days 1945
Eugenia Irene on Bank of Montreal steps rolling a cigarette with one hand

Eugenia Irene on stampede grounds

Eugenia Irene on the streets of Vancouver with violin and music

Anne

Eugenia Irene

Edward

Henry

Patrick

Marshall

Joyce

Paul, Anne, Dana, Keith

Eugenia Irene and son Billy

Henry, Joan with their eight
Mark, Matthew, Marshall, Lorraine,
Melvin, Michael, Laura, and Linda

Dianne, Edward, Lorne, Clinton, Suzanne

Patrick,Connie, Craig, Glen, Dean

Voya, Jo, Yuel, and Marshall

Curt, Joyce, Marie, and Rhonda

Chapter 24

Not until the opening of the Hope-Princeton highway in 1949, the Fraser Canyon was the only way to go. Hardy motorists travelling to Vancouver from Quesnel, the Cariboo; from Vernon and the Okanagan Valley, during the 1940's travelled the Fraser Canyon "The Old Cariboo Wagon Road". Today the Old Cariboo Wagon Road is but a memory. The Fraser Canyon road during the early 20's, 30's and 40's is unspeakable. It was extremely unfortunate that the final road width in the canyons was only sixteen feet (on sidehill sections it was even reduced to fifteen feet), a "hair-raising" experience for travellers. The same section of roadway, separated by a period of sixty years was no wider, and in many places it was narrower.

The original Cariboo Wagon Road from Yale to Barkerville, some 400 miles was one of the most difficult construction jobs, but the colony needed this road to the Cariboo gold fields. From 1861 to 1863 a small body of Royal Engineers sent from England surveyed and supervised the construction of this 400 miles road. Six miles of the Cariboo Wagon Road through the first part of the Fraser Canyon was their most difficult and hazardous project; their survey included Sapperton at New Westminster, Lytton, Lillooet and Quesnel.

Almost as if in protest, Hell's Gate erupted in a gigantic rock slide on March 6, 1914, altering the old course of the Fraser River forever and creating a real hazard for the ocean going salmon which was not solved until a special runway was built around the jumble of rocks in later years. By the time the First World War broke out in August 1914, the Fraser Canyon had been badly battered by the needs of commerce. Nevertheless, the Canyon had one consolation. By means of slides, flood, and rockfall, it had killed off the Great Wagon Road. No longer did the four-horse stages careen surely but with reckless skill over the trail, no longer did the ponderous freight wagons ply their trade through the defiles, no longer did the paddle wheel steamers breast the swift running currents. But, a new and greater menace to the Canyon's serenity appeared – The Model T Ford. World War I intervened slightly holding back the conquest of the motor vehicle. They were hard years which forced the consolidation of several rail lines. The picturesque Kettle Valley line was absorbed by the C.P.R., while the Canadian Northern and Grand Trunk Pacific amalgamated with the Canadian National Railroad. Here and there adventuresome motorists struggled up Jackass Mountain in a cloud of steam, but - like the Simon Frasers - they were more exploratory than commercial.

Construction of the first Fraser Canyon Highway started in 1924. The highway was completed to Lytton in 1926 and to Ashcroft the following year. In the mid-twenties, the Trans Canada Highway came off the drawing boards and began to be a reality. The Old Wagon Road up the Canyon was repaired and revamped; it was a good-weather road only – defined to mean "the best of all possible weathers". Nevertheless, by 1924-25 a passable road had been constructed from Vancouver to Revelstoke, the old bridge at Alexandra had been replaced by a modern trestle and by-passing Ashcroft, continued north to a point some six miles farther north. Here at a place called Cache Creek (so named from a legend that a robber disappeared into the muskeg near there with a reputed $70,000 in stolen gold which has never been recovered). The Trans Canada turned eastward while another pair of dust-ridden ruts headed north for Clinton and the Cariboo. Dust-proofed motorists stopped at Ashcroft Manor, as did all other travellers.

With the coming of the Depression, 1929, a poor-man's Gold rush into the Cariboo began. Though the Rush of 1860 had taken the champagne from the bars, there was still beer money and hundreds of men came on foot, by bicycle, on rafts, by car and by rail to grub in the sandbars above and below Quesnel. Though none made a fortune, many made a living – which was very important during the Depression.

The Second World War, September 1, 1939, brought no great improvement to the Fraser Canyon. If anything, it seemed to solidify the two railroads as master carriers. But with the 1950's, an infant trucking industry was born, and their 13 ton monsters began to challenge the grades and the turns of the Canyon. At Cache Creek, a modest café was built to accommodate the truckers. Gradually, road-side gas stations began to crop up along the way; and a program of redevelopment of the Old Wagon Road began. Saddle Rock tunnel, three and a half miles from Yale was started in spring 1957. Extensive work was done around the Nicomen River, and another tunnel completed at Sailor Bar. Hell's Gate was conquered by tunnel the following year and work was begun on the retaining wall around the great slide area 12 miles south of Boston Bar. The longest tunnel in the highway is the China Bar Tunnel at the China Bar Bluff, named for the old mining area on the Fraser River eight miles below Lytton. This tunnel is one of the longest highway tunnels on the North American continent. Construction began in 1960, and one of the main hazards was the Canadian National Railway mainline track

150 feet below the highway. The last tunnel to be built was Yale Tunnel completed in 1963.

Between 1950 and 1970 many bridges were built on the highway, including the new, gracefully arched Alexandra Bridge, just south of the old Alexandra Bridge still visible upriver from the new one. That was the third bridge at this crossing; the first the suspension bridge of 1863, and the second built when the original was destroyed by the 1883 flood, became obsolete and could not carry the increased traffic. Not until the opening of the Roger's Pass section of the Trans Canada Highway on July 31st, 1962, did the Canyon come into its own as a tourist attraction. Cache Creek blossomed, while its rival Ashcroft – by-passed by the modern ribbon of road – wilted. The former, tailored to meet the needs of a transient clientele contrasts keenly with the older town. While nothing but the growl of exhausts disturbs the modern town, sometimes when the wind is still, imagination can bring the sounds of wagon wheels crossing the Ashcroft bridge.

The present Fraser Canyon Highway is one of the most scenic stretches of the Trans Canada Highway. Travelling the narrow valley of the Fraser River, one sees the magnificence of the mountains, as well as the winding beauty of the river and the many sandbars where miners panned for gold. Yale is now a modern village and very proud of its historic heritage. Spuzzum consists of a gas station, store and a few dwellings. Boston Bar is more commercial with sawmills and logging. Kanaka Bar has a motel, restaurant and a magnificent view. Lytton still has a warm welcome. From a roadside pullout at Lytton can be seen the joining of the Thompson and the Fraser rivers.

Where the two great rivers meet; they keep on rolling;

Like old man river,
dat ol' man river,
He must know sumpin',
but don't say nothin',
He jus' keeps rollin',
He keeps on rollin' along,

Leaving but memories in the sands of time.

Family portrait September 17th, 1948

Chapter 25

Sentimental Journey

When my father brought his family to the Okanagan Valley from Alberta the summer of 1943, in his 1938 grey Plymouth car, the Big Bend was still a 15 and 14 foot wide wagon trail; and when he chose to take Edward and me along that same old trail, back to his old homestead, and for us to see once again our old farm and birth place in the autumn of 1949, there had been very little to no changes made to that Big Bend Trail; and from Banff to Revelstoke. The Fraser Canyon to Cache Creek, to Kamloops, Salmon Arm, Revelstoke to the Glacier National Park, the Big Bend to Golden to the Yoho National Park, Field, the Kicking Horse Pass, Lake Louise and Banff, to Calgary, Alberta and on to the Maritimes; for the first time became a part of the Trans Canada Highway with the passing of the Trans Canada Agreement on April 25, 1950, but due to many negotiations it was not finalized until the early 1970's.

Although the road had been graded in areas, there were many sections that were hazardous. Cliffhanging pieces of road had been made safer and better bridges installed, but the road had actually changed little. These highways and byways I have mentioned were a big part of my father's life, not only did he travel along those hair-raising trails constructed in the 1860's, then repaired slightly in 1924 -1926, he was helping to build the country roads that today are still being travelled on in Alberta, where we used to live.

Driving that route through the Glacier National Park, we travelled 50 miles north along the Columbia River to cross, then 50 miles back south and into Golden. That 100 mile bend is how the Big Bend got its name, which became obsolete when the Rogers Pass opened in 1962. My father, Edward and I travelled this old trail in the autumn of 1949, in a 1949 half ton Fargo truck. This old trail was gravel, had many holes and was not fit for travelling in a motor car. Father was aware of the hazards, from his drive through the Rockies with his family in 1943. For this drive through with Edward and me; he bought the Fargo truck. There were many old broken up nailed together wooden bridges to cross the Columbia River. Some nails were missing on those bridges and there were nails sticking up out of the loose wooden boards we had to travel along to cross the river. In the late autumn, the time of year we travelled; the Columbia River was low, very wide and with many sandbars. The sandbars were graded and made into a trail and the wooden bridges were built to cross the running river. We got a flat tire midway across the Columbia River where we had to stop on a sandbar, and while father jacked up the truck to pull the nails out of the tire, and to patch the inner tube, Edward and I strolled along the river's edge looking for and collecting pretty rocks. It was quiet there on the Columbia River sandbars, just the soft gently flowing sound of the river could be heard. The sun was shining brightly and the autumn leaves along the river's edge and on the mountain sides were beginning to show their pretty autumn colours. Father got this half ton Fargo truck fitted with a canopy to take his first born son Edward, with him, Mallard duck hunting on the lake on his mother's old homestead farm where he was growing up, and where he alone had trapped for beavers, to put himself through high school. He wanted me to be there too, with them, the three of us together who travelled these majestic mountains by rail just six years previous then as children with him in a boxcar; now as young adults in a Fargo truck.

Gonna take a sentimental journey
Gonna set my heart at ease
Gonna take a sentimental journey
To renew old memories.

Father wanted me to see my birthplace, and his old homestead where he was born, before I again left home. On our return from this trip, I was to travel to Toronto to study violin, harmony and theory at the Royal Conservatory of Music in Toronto. And I would be away from the family for many years. There were other reasons why father took Edward and me along to be with him and he was not telling us though today I know that it was for us to just be there. This would help him decide if he should sell, or if he should not sell the farm we were born on. This farm in 1949 was still very much a part of us, very overgrown with tall grass and looking very sad and empty, though still a big part of our lives. The six years we were away from this farm had passed quickly though had left their mark, on this our log cabin home and on this land we knew and loved so well. Being here again, quietly listening, I could hear the sounds and the echoes of our life, and our laughter all around me, through the trees, and all around our log cabin home as it used to be: like a spirit, like a soul, that lives on and on, and would be there forever.

Got my bag, I got my reservation,
Spent each dime I could afford.
Like a child in wild anticipation,
Long to hear that "All aboard".

Seven, that's the time we leave, at seven.
I'll be waiting up for heaven
Counting every mile of railroad track that takes me back.

Father's sister Julianna with her man Paul, together were still farming grandmother's old homestead and were looking after her first, and original, old pioneer shack. They had built a bigger and better house for their own family and they built on this farm more barns and sheds for all their animals, and for all their farm machines. This 1949 autumn trip with Edward and me was father's "Sentimental Journey Home".

Never thought my heart could be so yearny,
Why did I decide to roam?
Gonna take a sentimental journey,
Sentimental journey home.

Here on this land, the homestead where my father's parents first came to and settled in the 1800's, my father was on his birth farm, not as a little boy with possibly a step-father, or a brother–in-law at his side, but as a proud father with his own first born son at his side, and together duck hunting where he used to be and all alone, catching beavers. The wild Mallard ducks here on this homestead lake were grain fed, and all lived on this cool, clean, clear lake year after year, and there were many. Many, many of those fat healthy little Mallards on this homestead lake, that seemed so much smaller now, than it was when father was a little boy. Julianna and I stayed at the house plucking and singeing these ducks over an open outdoor campfire, cleaning and washing them while father and Edward kept bringing in more and more.

While Julianna and I were trying to keep up with the many that were being brought in, father and Edward went into Innisfree, and they brought back 12 cases of quart sized jars. There were 12 jars to each case, and we, Julianna and I filled them all with young Mallard duck. After we had 144 individual quart jars stuffed with duck, we then placed the filled jars into canning boilers, covered the jars with hot water and put them on to cook for three hours each boiler. Julianna did not have the space on her stove and she had two stoves cooking. One stove in her kitchen and she used the old stove in grandmother's pioneer shack. All the extra leftover jars of duck for cooking, we took to Julianna's daughter Eva's place in Ranfurly, where my mother with her brood were left waiting for father's return in June 1943, when father, Edward and I left the Ranfurly stockyard in a boxcar and on our way to our new way of life through the Rocky Mountains.

During the three hours waiting, for the jars to cook and to seal, my father with his son at his side went out over the fields to re-trace his footsteps over the hills and valleys of his homestead searching for a wooden cross. When they returned the work was all done; the ducks were cooked then cooled. We placed them all back into their cases and father carried them all out to the truck and placed the cases side by side into the back of the truck under the canopy and he covered them firmly with his furniture tarp and we were on our way. We stopped at my grandmother's cottage in Mundare for overnight before starting our long journey through the Rocky Mountains and on our way back home. Grandmother gave us two of her homegrown live geese in a crate to take with us to our Okanagan Valley home, and she packed a picnic basket of food for us to have along the way. We left grandmother's cottage, and Mundare, very early the next morning and by nightfall, we arrived in Field, British Columbia. We had just come down a very steep and very narrow mountainside to get into Field. The mountains here were very high and all around us. It was like being in a canyon. This was in the Yoho National Park. At 30 and 35 miles per hour – which was the speed limit during the 30's and 40's, we came in all the way from Mundare through Banff, Lake Louise, the Kicking Horse Pass and into Field. During this journey, we did not stop in Banff; we stopped in Lake Louise on our return and had our pictures taken in front of the Lake Louise train station; we had our picnic lunch here, and again were on our way. When we were stopped in Lake Louise, the sun was still quite high in the sky though the wind was blowing strong and very cold. We did not stay, but hurried on. My father was not going to drive the narrow mountain trails in the dark of night. In Field where we stopped, father got us a cabin for the night. There were no motels, as are motels of today, they were overnight little cabins, like trappers cabins. Father brought the crate with the two live geese into our cabin for the night; he said to keep the timber wolves from coming near. All through that night, the eerie, weird, lonely sounds that echoed, were very much the same as the sounds we heard in the dark of night on the 4th of July 1943 when we were stopped for the night in a boxcar, with our five cows, a team of horses, a crate of chickens, two pigs and a pair of ducks. This dark of night with our two geese in a crate, it seemed we had been here before.

We left Field at sun up and were around that Big Bend by daylight. With a heavy heart, father was returning to his Apple Valley home. He was a quiet man within. He did not say much to express his feelings, and his sentiments of when he was a little boy were private. He was thoughtful, emotional and kind. Not only did father want to take his first born son duck hunting with him on his homestead land, he took Edward and me with him on his Sentimental Journey, in search of that "Old Wooden Cross" somewhere on a little hill, under a favourite tree is where his father Michael was buried on this land, his homestead in 1905. He died as a young man while building the Canadian Railways, for trains that no longer run. With his son at his side, father walked the fields searching on every little hill under every tree left standing for that old wooden cross marking a grave so dear to his heart; a grave of his father whom he never knew though cherished the memory of through the years. And had hoped to someday find "That Old Wooden Cross".

On a hill far away stood an old wooden cross,
The emblem of suffering and pain:

And I love that old cross where the dearest and best
For this world a lost dear one was lain.

So, I'll cherish the old wooden cross.
Till my trophies at last I lay down;

I will cling to the old wooden cross,
And exchange it some day for a crown.

My father never could exchange that old wooden cross for a crown:

It was never found.

Our last visit to our farm and home in autumn 1949 Eugenia Irene and Edward by the front door

Paul, Juliana, William my father, during duck hunting

Grandmother's first homestead. Father walked with his son Edward searching for the old wooden cross. It was never found. 1993 - William's baby daughter Joyce with Julianna's baby son Harry who still lives here.

Chapter 26

"The Homecoming"

Driving through the Rocky Mountains from Banff, to Field, Golden, around the Big Bend and on through the Monashee Mountains, Revelstoke to Vernon along a bumpy gravel road with many pot-holes and ruts, we returned to our Apple Valley home. Father seemed a little bit sad, as if a part of him was left out there on his old homestead still searching for that old wooden cross; though he was pleased that we brought back with us a supply of wild duck from his homestead lake, cooked and ready for serving during the coming winter months.

He felt guilty that he had left his Katharine alone with four young children and completely alone in charge of the B-Y Store, with no extra outside help while he went away, a going on and on, searching. His determination to find his father's grave forced him to go on. And now, 44 years later, late in the month of September 1949, to no avail his searching had come to an end, leaving a void, An Elusive Dream.

I know you're tired of following
my elusive dreams and schemes
For they're only fleeting things'
my elusive dreams

The two live geese we brought with us from my grandmother's home in Mundare, father let them out of the cage into the chicken pen to be with our chickens and he brought into the house all the cases of cooked wild duck. As father was bringing the cases of duck into our house; I was telling my mother of the work and the effort that Julianna, Eva and I put into preparing and preserving the 144 jars of duck meat for her to have ready for serving to her guests and to her family during the coming winter months, and together, my mother and I went into our kitchen to unpack the jars from the cases and to store them away in the pantry - - - **every jar we took out of its case was unsealed.** The bumpy ride in the back of the pickup truck unsealed every jar, though the jars were all well packed and covered to keep them from bouncing. What a disappointment; and after all our work and effort. We emptied every jar of young Mallard duck meat into big buckets and father hauled them all away to the city dump. The Vernon city dump always had a fire burning. All the fruit the packing houses could not sell at the set summer prices was hauled away to the city dump; gasoline was poured over the fruit, then set on fire. We could see the fruit bonfire from all parts of the city as it was burning for days and nights, topped off with our Alberta homestead Mallard ducks. We were sad - and we were speechless - and I do not know if my father ever told his sister Julianna what had happened to our Mallard ducks that she worked so very hard to help us prepare and preserve. After all was done, I was pleased that father did take Edward and me along with him on his "Sentimental Journey". The drive along that wagon trail through the mountains in a pickup truck, was interesting and exciting. Most of all just being there with my father when, I believe, he really needed us to be there with him is most gratifying.

Mother wanted a house away from the B-Y Store she said, a house with a big yard and a view, and she asked me to go house hunting with her. We set out on foot, she and I, to find what homes were available and where; and that very same day we found one, a brand new home a German builder had just finished building. The address was 4212 – 33rd Street, Vernon, near the junction to the Kamloops highway. Across from this house were no buildings along 33rd Street; the land was cleared and was as large as a softball field, with Swan Lake Creek running through at one end along 32nd Street and on through the Vernon Lodge on its way to the Okanagan Lake. From the front porch of this house we could hear the relaxing, babbling sounds of the creek, and we could see the apple orchards across on the other side of the valley beneath the Silver Star ski mountains. In the yard was a vegetable garden, a chicken pen and a garage for our car. Mother and I hurried back to the B-Y Store and we told father about this house we found, and while we stayed and looked after our customers at the store, father went out and he bought this house from the builder, and in a few days the family moved in.

This house had a full concrete basement, two bedrooms on the main floor, the master bedroom was large. There was a large living room, dining room, kitchen and a bathroom. The attic was finished as one large room though there was no staircase leading up to the attic. Father built a ladder in the second bedroom clothes closet as a temporary entrance to the attic for Edward, Henry, Patrick and Marshall. This attic would be their private room to sleep in, and for doing their school homework in, and father fastened coat hooks along the wall across from the closet door for them to hang their clothes, their coats and their jackets on, before going up the ladder. Joyce would have the bedroom.

Anne was coming home to see our new home and to visit with the family. A Vancouver City Police Officer brought Anne home, he was her friend and his curiosity of meeting her family encouraged him to drive the Fraser Canyon to bring her home. His name was Bob. Everyone were at their very best behaviour knowing a Police Officer was amongst us, and we gave him the bedroom as his private room for the night. Joyce was seven years old; for her we set a cot in the master bedroom for that night, the boys had the attic and Anne and I slept in the living room in sleeping bags. The four boys went to bed, up into the attic early that evening, they had school to go to in the morning; so did Joyce, and she went to sleep on her cot, while father, mother, Anne and I stayed up quite late into the night getting acquainted with Bob. Very early the following morning Edward, Henry, Patrick and Marshall quietly, one after the other, came out of that closet in Bob's bedroom not to disturb the sleeping Police Officer; to have their breakfast and to be off to school. Bob was going to stay put, in bed, until after the rush to the bathroom would be over with and after everyone was calm and quiet and sitting at the breakfast table before he was going to join the commotion. With one eye open, seeing four boys, one after the other, coming out of that closet in his bedroom was amazing, he had never ever seen such a thing; and his curiosity got the best of him. He came out of that bedroom looking stunned and the look on his face was unbelievable. He said he got up and he looked into that closet and all he saw were cloak hooks on the wall and he was wondering if the four boys hung themselves on those cloak hooks for the night, or did they all sleep standing up jammed in together in that closet; it was a very small closet. The year 1949, coming out of the closet had a much different meaning from what coming out of the closet means today.

We all had breakfast together seated around our dining room table, and with a few laughs, then our good-byes; the students got on their school bus and were off to school. Bob left that same morning; he had to be back on his job protecting people on the streets of Vancouver. His visit with us was short but interesting, and we all enjoyed having him with us for a day. Mother, Anne and I went into town to the B-Y Store to be with father and to clear out what items might have been left in the living area of the store after the moving truck took most of the clothes and furniture to our new house. There were a few things left for sorting, and after everything was removed from the back of the store, that whole living area was used for storage. Father was bringing in more stock at one time and he did not have to travel to Vancouver as often as he did when there was no space to store his retail stock in.

Many evenings before sundown, on the empty lots across from our house on 4212 – 33rd Street, a softball game was being played. Both the little children, and the adults loved to play softball, and there on those empty lots, they set up a softball diamond, along the Swan Lake Creek. Our family had a white Bulldog-Terrier dog with a black tip on his tail. We called him Tipy. He too played softball, and was the best catcher on the team. One day when just the little children were playing, the ball went into the Swan Lake Creek, and as Joyce ran to fetch the ball;, Tipy ran after her and he pulled her back away from that creek by the seat of her pants, far enough away, he then with a swift leap and a jump was fighting the fast running water and he fetched the ball, swam back to shore; and delivered the ball to the pitcher. He was quite the little dog; really tough and an important member of their softball team.

Henry, at the age of 15 years, was an inventor; he was building real rocket ships with real gun powder. One day he just about blew away his whole thumb off his right hand, and our family doctor, Dr. Campbell-Brown was always there, he often made house-calls because he cared; and he was there to mend all scrapes and bruises. He was a super-special doctor. He encouraged Henry to continue with all of his inventions, and he was always very proud of him; Henry was his number one.

Edward was more of a merchant, he was relied on in the store and all our customers liked Edward, his pleasant smile and his eagerness to help. Patrick Marshall and Joyce looked after our chickens, cats, dog and the garden. And mother walked from this new house the 12 blocks to the B-Y store every day before noon. She carried father his lunch and every day she was there, at the store to have lunch with father.

The neighbours on either side of us on 4212-33rd Street, admired our family; they said the children were all very well raised. There were the Sturgess' on our right sharing a driveway with no fences in between. They had one daughter who was 8 years older than Joyce. On our left were the Lovelace's, they had a young daughter, and one son older. And ! never did our cats, dog, chickens or children ever set foot on either of the neighbours' properties without first being invited. Both neighbours were pleased and grateful with our concerns and remained special friends of our family all through their lives.

The last day of October 1949 had come and I was to be on the road again. Early that Monday morning October 31st I went to the B-Y Store with my father to reminisce and to be with him before I was going away. I had my violin with me though not many clothes. I did not have many clothes so there were not many to pack. Just before noon that day a young man walked into the B-Y Store and was asking my father's permission if he may take me out to the restaurant next door for a soda before I leave town; my father said that I was there and that he could ask me. He was from my class during my high school days at the Polson Park High School. As we sat at the counter in that restaurant sipping our sodas, he said to me that he did not want me to go to Toronto, he had marriage on his mind. That remark took me by surprise; I did not really know him all that well and I told him that my music career was set, with no turning back. His name was Silver. He was one of the boys who were at the Vernon Stock Yard the day our box car landed, towed in behind "The Galloping Goose", with my father, Edward and me in it.

I returned to the B-Y Store and found that my transportation to Toronto was there waiting for me. My transportation to cross Canada was a camper truck with a music teacher and his wife. Originally they were from Hamilton, Ontario and after teaching music in the sunny Okanagan Valley for a few years, they were returning to Hamilton and offered to take me with them to Toronto. They told my father that they did not want money for my transportation, they just wanted me along, to be camping and travelling with them as their companion. They were a middle-aged couple, in their early 50's. His name was Andrew and her name was Irene. Father gave Andrew $100.00 for my expense he said, and he insisted that Andrew take it. $100.00 was a lot of money in 1949, and for that, Andrew was to find me a place to stay in, when we reach Toronto. I had no money when I was leaving, not a penny to my name and I never would have taken money from my father. I had my completed Business Course to rely on. I gave my father, mother and Anne a big hug and I cried a little, then boarded that camper truck and we, Andrew, his wife Irene and I drove away, to where, and how, was yet to come. All the way across Canada, Andrew had two Irene's in the camper truck with him,

and every day,

all the way,

he sang:

"Good-Night Irene"

Goodnight Irene, goodnight Irene, I'll see you in my dreams.

Last Saturday night I got married, me and my wife settled down
Now me and my wife are parted, I think I'll take a little stroll downtown.

Sometimes I live in the country, sometimes I live in the town and
Sometimes I take a great notion, to jump into the river and drown.

Stop your ramblin', stop your gamblin',
Stop staying out late at night.
Go home to your wife and family, and sit by your fireside bright.

Irene goodnight, Irene, Irene goodnight.
Goodnight Irene, goodnight Irene, I'll see you in my dreams.

Chapter 27

The Hope-Princeton Highway was unveiled on November 2, 1949, and father made his last trip of the season, for the first time, along that new highway. He took his daughter Anne with him; she had her job at Woodwards to get back to, and together along that new paved highway they were enjoying some of the most dramatic scenery in Western Canada. At times this highway courses between valley walls lush with vegetation and history, and at others it climbs mountain sides to meander gracefully over the top of the world. From that day on, November 2, 1949, that new Hope-Princeton Highway became the only route my father travelled along year after year, and sometimes several times during a year for goods to stock his store with for his customers, their families and for their homes; his customers who lived all through his beautiful Apple Valley from Salmon Arm, Armstrong and on through the valley to Penticton.

In Vancouver father loaded his truck to capacity for his return trips along that new paved Hope-Princeton Highway, there were no more ruts, pot-holes and wooden bridges to cross. He said it was smooth sailing all the way. And since the family all moved to their new home at 4212 - 33rd Street, father had ample room to store all his goods in to last for six months and more.

This new Hope-Princeton Highway was originally the Dewdney Trail and it ran from Hope, through Rock Creek and on to Wild Horse Camp, six miles north-east of Cranbrook. A 24 year old Edgar Dewdney, a civil engineer from Devonshire,

England who had produced excellent work in the survey of the city of New Westminster got the contract, and in the spring of 1860 went to work on the first leg of what was to become the famous Dewdney Trail. Edgar Dewdney's program was to construct a road 4 feet wide from Hope to Rock Creek, but in reality only the centre two feet was guaranteed to hold beast or man. At one point it rose to the breath taking height of 6,000 feet, creating one of the highest passes in Canadian highway history.

The original Dewdney Trail followed the present route of Highway No. 3 from Hope to the mouth of the Skagit River. At this point it looped north from the modern highway and then east to the headwaters of Whipsaw Creek and followed this torturous river, to a spot called Vermilion Forks, now Princeton. The town of Princeton, British Columbia was incorporated in 1951 and is located at the junction of the Tulameen and Similkameen Rivers. The location was called, Red Earth Forks or Vermilion Forks until 1860. It was then called Prince's Town, (hence Princeton) after the Prince of Wales (later Edward VII) who visited Canada in that year.

From Hedley to Keremeos the Dewdney Trail followed the present course of Highway No. 3, but when it left Keremeos it bent south and east to Osoyoos. Beyond Osoyoos, the trail had the formidable task of climbing Anarchist Mountain – a feat that defied many a Model T in the early 1920's. Achieving the top, it wound across the plateau – like walking through clouds on the top of the world – before it slid down the eastern slope of the mountain and coasted into the little mining community of Rock Creek.

For some time the Hudson's Bay Company had been pushing into the southern interior of British Columbia and in 1860, the completion of the original section of the Dewdney

Trail, the Company moved its operations to the site where Keremeos now stands. Keremeos, aptly, means "wind passage through the hills". The first factor, Francis Deachinquette, built a log trading post on the famous trail and started the first farm in the district.

Frank Richter, after whom the Richter Pass is named, came over the trail in 1860, establishing himself near Keremeos as the first fruit farmer which at one time extended from Keremeos to Princeton.

1861 the Dewdney Trail was completed to Rock Creek and from Rock Creek to where Midway is presently situated. Dewdney and his men struck the Kettle River, which was later to be made famous in connection with the first railway through the south. Most of the early travellers over the Dewdney Trail were prospectors. These found their way into the area through the Crow's Nest Pass from southern Alberta. At Grand Forks, the first permanent settlers were James McConnell, Robert R. Gilpin and William S. Jones who were open-range ranching. From the United States, the Corvette family arrived in 1885 and developed the first prune orchards in the valley.

The Canadian Pacific Railway construction crews entered the Crow's Nest Pass in 1897 and pushed through Fernie, Cranbrook, Creston and on to Kootenay Landing by 1898. In the fall of 1905, the Great Northern Railway from the United States, which had been enviously watching freight revenues climbing in the Boundary Country, obtained a charter to build into the district. With amazing speed, the Great Northern threw down rails from Spokane to Midway – extending spur lines up to Grand Forks and Phoenix. At Phoenix, 4,630 feet above sea level, the citizens not only boasted that they had the highest city in the nation, but also the highest judge. Magistrate Willie stood 6 feet 8 inches tall.

The Great Northern pushed its lines past Midway, zig-zagging across the American border where necessary, and reached the town of Oroville, just across the border from Osoyoos, in the fall of 1905. In 1908 strengthening its grip on southern British Columbia, the Great Northern Railway received permission to extend its line from Oroville – where it had arrived in 1905 – to Princeton, and on December 23, 1909, sent its first passenger train rolling into Princeton. Its coming spelled the end of W.E. Welby's little stage coach service. Later in 1912, this line was extended up past the coal mining communities of Coalmont and Tulameen.

Still – the Canadian Pacific Railway lines were tied in a knot at Midway.

On July 5, 1910, the first sod was broken at Merritt, and the Kettle Valley Line began. It was decided to take the tracks from Midway - where they had poised since 1899 - to the famous old gold town of Rock Creek. Here, instead of following the Dewdney Trail, the line looped far north through the mountains and approached Penticton from the northeast. From Penticton, the line would again loop north, through Summerland and Osprey Lake and come into Princeton from the northeast. By the end of 1912, 206 miles of track were laid and the line was completed to Princeton.

The second phase of the Kettle Valley Line was completed by July, 1916, when arrangements were made to cut off at Brookmere (north-west of Princeton) and run down the Coquihalla Valley to Hope. It was in this latter form – Midway to Brookmere to Hope – that the line operated for nearly 43 years. Though it continued to be called the Kettle Valley Line, the operation had been taken over by the Canadian Pacific Railway as early as 1913.

In the early days of British Columbia, the land lying along

the American border from Rock Creek to Salmo was almost forgotten territory. In the beginning, the fur trade followed southward by valley and river to the United States soil. With the discovery of gold, copper and silver in the Boundary country, a subtle struggle between American and Canadian influences developed – each striving to draw the trade from the area. Over the years, the history of the region has been woven around the struggle between the powerful American magnet of roads and railroads to draw the Boundary country into its orbit and the Canadian efforts to divert this traffic into an east-west pattern. The two major weapons in the hands of the Canadians were the Dewdney Trail of 1865 and the Kettle Valley Railroad.

On November 23, 1959, a washout occurred on the Kettle Valley Line near Jessica on the Coquihalla section of the line. Rather than face the expensive task of repairing and maintaining the Coquihalla section, the Canadian Pacific Railway petitioned and received permission to close this part of the line. On January 17, 1964, the last passenger train over the Kettle Valley Line left Penticton with two old-time railway men in the cab. James J. Warren, whose ingenuity had brought the line into existence, served as fireman, while Andrew McCulloch, the engineer who surveyed the route in 1910, acted as engineer. As the train rolled through Naramata, McCulloch, Rock Creek and Midway, old timers who could almost remember the first shriek of the first whistle, climbed aboard for the historic run. When the train pulled into Nelson that night, it rang down the curtain on nearly 50 years of service.

In the late 1950's and continued on into the 1960's the Old Dewdney Trail was being widened and strengthened in an on-going program and on August 15, 1964 the final link in the southern highway system was officially opened.

Unlike the Old Dewdney Trail from Hope to Princeton

which went north from Skagit River, the new Trail looped far to the south and then went almost due north through Allison Pass at 4,380 feet above sea level to Princeton. As it approached Princeton, it ran down a long tongue of land, underneath which a tunnel gave passage to the rumbling trains of the Kettle Valley Line.

The few remaining traces of the Old Dewdney Trail, Hope to Princeton, tell an eloquent story of slowly eroding mountains. For over 100 years small slides have come down obliterating portions of the trail and on the morning of January 9, 1965 at 6:58 a.m., came the ""Hope Killer Slide". An earthquake shook Johnson's Peak and an estimated 100,000,000 tons of rock, earth and trees began to slide down into the valley a mile west of Tashme near Hope. Reaching the valley floor it wiped out beautiful Outram Lake and raced 500 feet up the face of Mount Coulter before falling back spreading out like a giant "W", engulfing a yellow convertible carrying Mary Kalmakoff, Bernie Beck and Denis Arlit. A hay truck driven by Thomas Starchuk, and an Arrow Transport Oil tanker with Norman Stephanishin at the wheel. The Hope Killer Slide changed Highway No. 3 forever, and that same year, 1965 became the end of an era for my father, 1905 - 1965 – 60 years of hard labour since he was a little boy, took their toll. No more was he able to service his customers, no longer could his weary muscles and bones take the strain of heavy lifting and hauling. He was selling out. All the stock he had accumulated at the B-Y Store through the years 1945 – 1965; and the B-Y Block was up for sale.

Twenty years of buying and selling antiques, furniture and appliances, hockey equipment, boots and shoes, making deliveries, driving a big truck to Vancouver for items to stock the store with and never missing an auction. Like the last passenger train over the Kettle Valley Line; this was the end of the line for my father and his B-Y Store. Of his seven children, not one of them was interested in carrying on with the B-Y Store business. His daughter Joyce was 23 years old and flying all over the world arranging tourist flights for Air Canada. Marshall was at the University of British Columbia, and playing hockey with the U.B.C. hockey team. Patrick stayed in Vernon and was barbering at the Mission Hill Army Camp. He was always there for father. Alone he took over selling the B-Y Store stock to ease father's pain, from him being there, to see it all come to an end. Later Patrick became the Okanagan Valley Property Tax Assessor stationed at the Vernon City Hall. Henry was in Ontario, a Pharmaceutical Representative. Edward was teaching schools in Kamloops. Anne was married to Paul Chauvel and raising her two sons, and I, Eugenie Irene with a severe walking disability since 1960 was creating an Accounting Practice (Pacific DataSystems Ltd.) raising my son and providing a home for Peter Norton a C.G.A. and R.I.A. student.

Father alone now with his Katherine, sold their house at 4212-33rd Street, and purchased a little three bedroom house at 1705-33rd Street on Mission Hill across 32nd Street from the Vernon Jubilee Hospital overlooking the city of Vernon. He built on an attached carport to this little house for his new blue-grey Mercury car; and he purchased a 2 1/2acre parcel of land fronting Okanagan Avenue. This land looked towards the Okanagan Lake and the Kinsmen Beach and it sloped down to the Vernon Creek which ran through along the bottom end of these acres from Polson Park on its way to the Okanagan Lake. This 2 1/2 acres was a beautiful spot where the family often gathered for picnics; marshmallow and wiener roasts over an open campfire; fishing in the creek for trout and dancing and singing around the campfire.

My father was suffering a great loss when he had to let go of the B-Y Store and all his customers whom he enjoyed seeing, serving and conversation with; though his family were his life; his world. His love for his family through the years, his main goal, was to draw the family together, to be with him and to have fun. To care for each other and to be together as often as was possible. When we all could not meet in Vernon for a picnic on his 2 1/2 acres, or in Ellison Park for an early morning pancake breakfast then an all day swimming, boating and hiking; we met in Manning Park off the Hope-Princeton Highway, or on the shores of the Similkameen River near Princeton. In Penticton or in Kamloops. Every weekend, every holiday, every chance we got we were always together. Within the mountains and the hills of Vernon and Kamloops, father took us on picnics, swimming and fishing at Monte Lake, Pillar Lake, Bolean Lake, Spanish Lake, and to an Indian Reservation Lake near Falkland off the old Vernon-Kamloops Highway. We all got together for family graduation and birthday parties and every five years, father and mother celebrated a wedding anniversary. And not only did they invite their immediate families; they invited all of their relatives from father's side and from mother's side from all over Canada and the United States; to meet, to get re-acquainted with each other and to be together to dine, to dance, and to have fun. Most of their wedding anniversary parties were held at the Vernon Lodge Hotel where most of their guests had reservations to stay when they arrived, and where they were all catered to. Father continued bringing the families together through the years and on June 26, 1992 was his last. This day was his and his Katharine's 66th Wedding Anniversary. And from that day, June 26, 1992, like the Hope Killer Slide, our family togetherness shattered. The one strong link in our family chain was broken.

Through the years what my father did for his family, his kin-folk, and what he did for others will forever remain…

Dear To My Heart.

Mother and father with first grandson William

Father and B-Y truck with tarp cover

We're

SELLING OUT!

ALL STOCK MUST GO

including: • Chrome Suites • Lounge • Dressers • Beds (double & single) • Heaters (coal, wood & oil)

Plus many more items too numerous to list here. Come in; browse and BUY at tremendous savings

1966

B-Y STORE

3308 30th Avenue Phone 542-2833

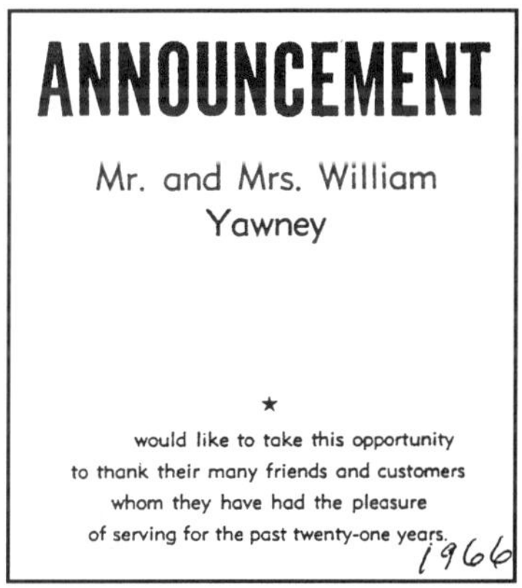

ANNOUNCEMENT

Mr. and Mrs. William Yawney

★

would like to take this opportunity to thank their many friends and customers whom they have had the pleasure of serving for the past twenty-one years.

1966

B-Y store envelope

Last Chance for BIG SAVINGS!

We Are Closing Our Doors on February 12

B-Y STORE

3008 30th St. - Vernon

Chapter 28

The B-Y 3-ton Chevrolet truck made many a miles travelling the Hope-Princeton Highway, and made many a deliveries to all father's B-Y customers. This Chevrolet truck - like a rock – served my father very well during his 20 years of hauling and delivering furniture and appliances and now was being put on an auction block to be sold to a highest bidder by his dear friend and business associate of 20 years in business; Vernon Auctioneer, Stan Hunt.

Father had a fastened tarp over the top part of this truck which made the appearance of a covered wagon. And before the truck was to be sold, he loaded it with a tent, camping and fishing equipment, a gas lamp, lawn chairs and camp stove, in the Autumn of 1965; he was taking his first born grandson (my son) William Alexander Frederick camping and fishing to the Cariboo and to Barkerville. When the truck was loaded with all the camping gear and equipment and he was ready to go; he always called out "All Aboard". That camping trip was to be the last run for the B-Y truck. All through his life, father was attached to railways and trains; perhaps the memory of his father who's life ended when the running of trains began stayed within him through all of his life, and the call "All Aboard".

My father loved his first born grandson with pride and compassion. His grandson, like he, was growing up without his father, experiencing a void that the two of them; grandfather and grandson, alone could share.

(One must live it – to understand it)

Their unique understanding of each other brought them closer together through the years.

In Barkerville the wind blew strong, then came a sudden, cold rain with a flurry and sent every camper up onto the back of the truck. With the gas lamp lit and hanging from the ceiling provided the campers with light and some heat; while on the camp stove some little fishes were sizzling.

Little Fishes
Little fishes in a brook
Grandpa caught them on a hook
Grandma fried them in a pan
Billy ate them like a man.

When the rain showed no signs of stopping, they all spread their foam mattresses amongst the lawn chairs and all night long as the wind howled and the rain swished, swirled and pitter-pattered on the canvas roof, they all comfortably slept in the back of that truck till the morning sun appeared and dried up all the rain. That camping trip was a cozy farewell to the B-Y truck and through the years fondly remembered by father's grandson William.

Since the B-Y Block was sold, Patrick made comments that father was amiss for not having a B-Y truck with a hydraulic lift through the years, a more modern way of hauling heavy furniture and appliances to save his own strength and the strength of the men who helped him with the loading and unloading. He said father was old-fashioned, that he did not approve of modern equipment.

Patrick's comments set wheels in motion and the search began through using the Internet and sending requests by facsimile to all Hydraulic Tool Manufacturers and to Truck

Manufacturers in the United States. Failing replies sent a search on for answers to Universities and to Colleges. And at the North Island College, Comox Valley Campus: On Fluid Power Industrial Hydraulics; and Mobile Hydraulics, Instructor: Ed Nelson had the answers:

1 Hydraulic lifts were available on a very few Army trucks only during the late 1940's.

2 Hydraulic lifts on all other trucks were not available, and were not recommended.

3 Hydraulic lifts were available by special order only.

4 Hydraulic lifts were not suitable for heavy lifting; would get stuck and slant out of shape.

5 Hydraulic lifts were not reliable: Special Orders were soon discontinued.

After all the searching; the answers were quite clear why my father did not have a Hydraulic Lift on his B-Y truck. A reliable ramp and a dolly were being used for loading and unloading during the 1940's and are being used today.

Summer of 1965, Patrick took charge of the B-Y Store. He was selling out – all B-Y Store stock. And he kept all the records to close the business for Mr. Atkinson; father's Chartered Accountant, to complete. For his own living expenses Patrick played a "Stand-Up Bass" with the Vikings Society Music Band in Vernon, and many other music bands all through the Okanagan Valley. Every weekend, every evening, he played for weddings, anniversaries, birthdays and at the Legions till he wore the skin and flesh off a finger on his right hand. With father and mother away my son William went to stay with Patrick, to help him tie his shoes, to do up his shirt buttons and to help him at the store. William was eleven years old at that time and together they both shopped, cooked and were brothers: William named Patrick "brother Gene" and that name stayed with Patrick even today. Father took that time-out to travel the new Dewdney Trail along the Boundary country to Trail. Many times my father travelled the old, then the New Dewdney Trail to Trail during the years 1945 – 1965. His Katharine had a cousin in Trail; her father's only relative in Canada, Alec Orenchuk, whom her father Metro Orenchuk sent for and paid for his passage to Canada during the early 1900's. After Alec arrived in Canada he found his way through the Crows Nest Pass to Trail, and there in Trail is where he stayed. He met his bride Annie in Trail where they were married and together through the years, they raised a son Bill and a daughter Jenny.

During his life in Trail, Alec Orenchuk worked at the Consolidated Mining and Smelting Company in Trail. And after 50 years of excellent service, he received a gold watch. No pension; a gold watch.

My father loved British Columbia with a passion, and because he was emotionally attached to railways and trains, during the 1940's and 1950's he often took his Katharine on romantic trips by train along the Kettle Valley Railway, going through and over the mountains to Trail where his Katharine could visit her cousin and his family. Father said that travelling the Kettle Valley Railway line was particularly beautiful and together they could relax and enjoy the scenery of Beautiful British Columbia.

The year 1965 they travelled by car to Trail, then on to Creston, Cranbrook, Invermere to Radium Hot Springs where they stayed for a few days to a week. From Radium Hot Springs they travelled to Golden, Revelstoke, Armstrong to Vernon. Or, more often than not, they returned along the Dewdney Trail to

Trail where they could rest and relax before going on and over the exciting hair raising cliffs and curves of the Dewdney Trail from Trail on to Osoyoos to Penticton, then north from Penticton along the Okanagan Valley Highway to Vernon: Their Apple Valley Home.

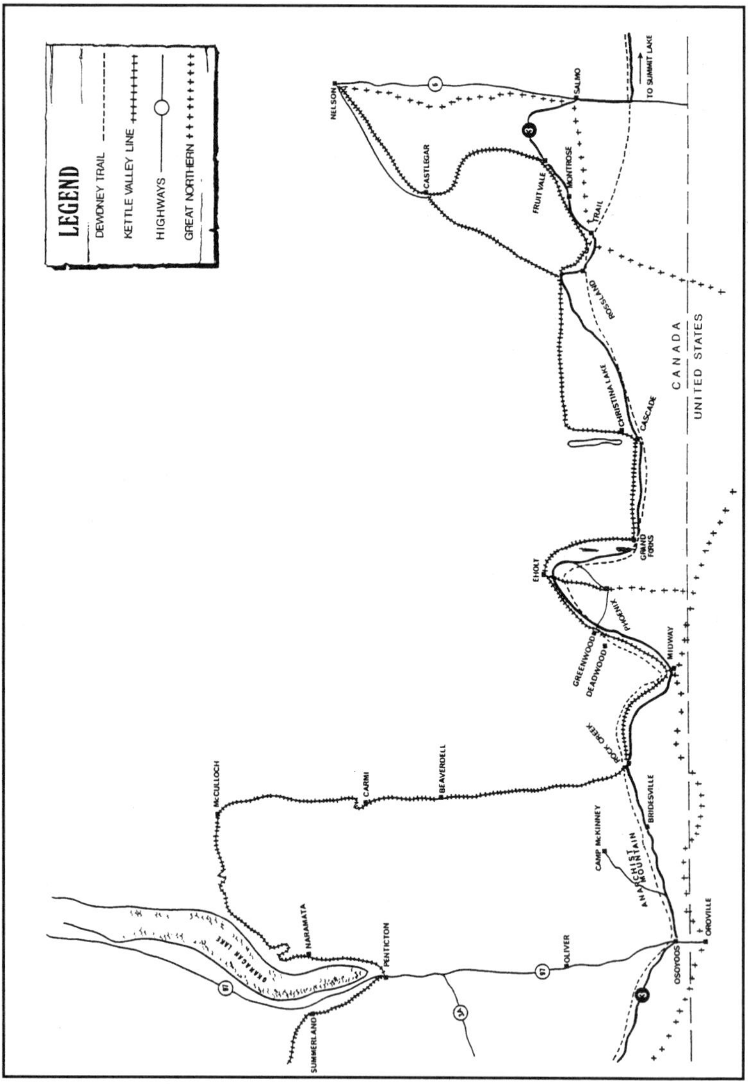

In the spring of 1890 a Deputy Mine Recorder, Eugene Topping, liked the ore samples on Red Mountain. These were the War Eagle, the Center Star, The Idaho, and Virginia. Thus, a fifth claim – The Le Roi – was staked adjacent to the Center Star and registered in the name of Eugene Topping. The next year, 1891, the claim showed good promise and a narrow crude trail was cut from the mine down to link up with the Dewdney Trail. The junction was right on the 320 acres of property pre-empted by an enterprising young man named Ross Thompson (the site of present Rossland). From here the Dewdney Trail was followed along Trail Creek for 7 miles to the confluence of the creek with the Columbia River. Eugene Topping, together with a man named Frank Hanna built cabins at the mouth of Trail Creek and they called the site – Trail – after the famous old trans-Montane route.

A disastrous flood in 1894 wiped out the little settlement of Trail, washing away all the buildings except the shed in which gold ore was stored. It was a good omen! A new Trail rose on the mud of the old and within a year nearly 2,000 people were living on the slopes of Red Mountain.

In 1895, a capitalist from Bute, Montana, F. Augustus Heintz, moved into the area and under him the economic structure boomed. Work was started on a smelter at Trail in 1895, and on February 1, 1896 it produced its first ladle of gold.

In 1898 the twin–settlements of Rossland and Trail were linked up with the Crows Nest Pass line and the Canadian Pacific Railway.

During the next few years, most of the smaller operations on Red Mountain either folded or were absorbed by the three giants – Le Roi, Center Star and War Eagle – and by 1905 the Center Star and War Eagle amalgamated to form the nucleus of the Consolidated Mining and Smelting Company. When in 1911 the Le Roi holdings were added, Consolidated Mining and Smelting emerged as the new giant in the mining industry.

With the opening of the Hope-Princeton section of the new Dewdney Trail, an increasingly heavy stream of traffic began to flow eastward through Osoyoos, Midway, Greenwood and as far as Grand Forks.

The cautious tourist turned southward at Grand Forks and into the United States rather than hazard the breathtaking curves and cliffs of the gravel road between Grand Forks and Trail. The venturesome motorist who did travel the Cascade Highway between Grand Forks and Trail looked upon the feat as a major achievement of their trip. A second hair-raising trip along the coastal highway went into Creston. From Creston, the highway followed the old Dewdney Trail to Cranbrook and on to Invermere and Radium Hot Springs along the Columbia River.

From Grand Forks to Trail, a new highway – almost paralleling the old Dewdney Trail – was opened on October 5, 1962, eliminating the breathtakingly beautiful but scary Cascade Highway. At Trail, the new Dewdney Trail took leave of the old by looping northward through Montrose, Fruitvale and Salmo, but at Salmo the highway drives south to join the historic route and wind hand in hand up the valley to Summit Creek. The Salmo – Creston cut-off, officially opened on August 15, 1964, was the final link in the southern highway system.

And with that official ceremony, the Dewdney Trail was again complete.

The new Dewdney Trail avoided the Richter Pass and circled it to the north. It was not until July 7, 1965, that the Richter Pass highway was opened and once again the old and the new highways crossed the mountains hand in hand.

With the opening of the Richter Pass highway, tourists were

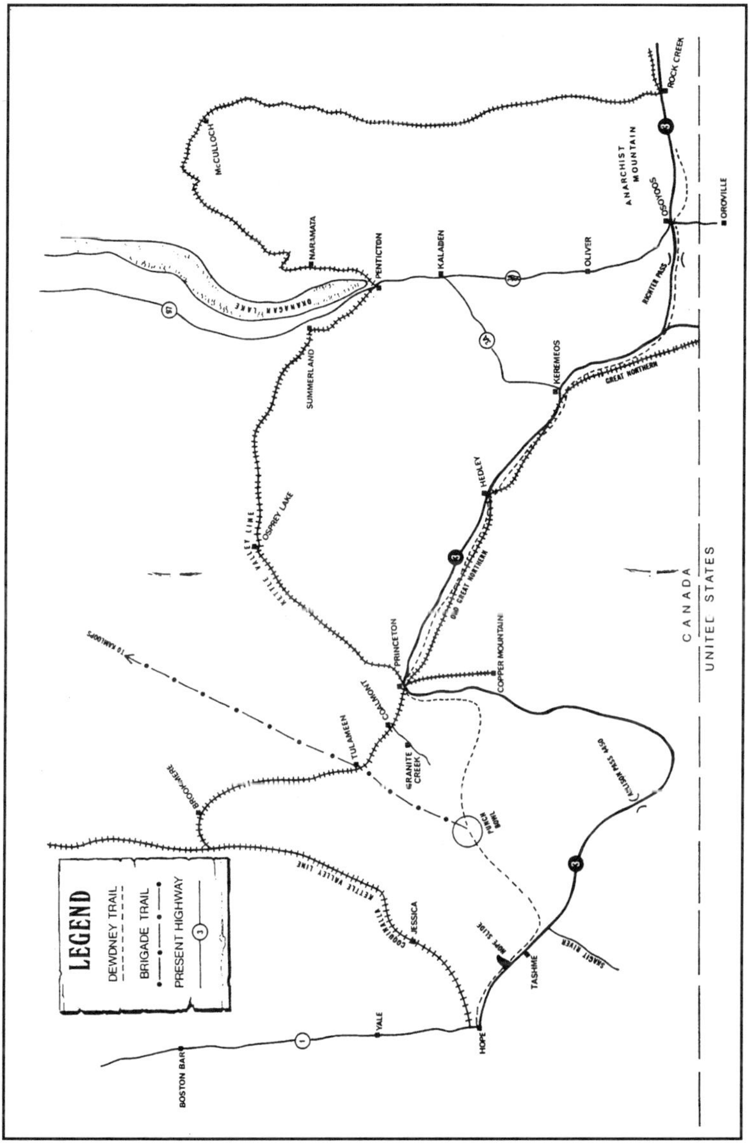
LEGEND
DEWDNEY TRAIL
BRIGADE TRAIL
PRESENT HIGHWAY
BOSTON BAR
YALE
HOPE
HOPE SLIDE
TASHME
SKAGIT RIVER
JESSICA
KETTLE VALLEY LINE
BROOKMERE
TULAMEEN
COALMONT
GRANITE CREEK
PUNCH BOWL
PRINCETON
COPPER MOUNTAIN
TO KAMLOOPS
OSPREY LAKE
HEDLEY
OLD GREAT NORTHERN
SUMMERLAND
OKANAGAN LAKE
PENTICTON
NARAMATA
KALADEN
KEREMEOS
GREAT NORTHERN
OLIVER
RICHTER PASS
OSOYOOS
ANARCHIST MOUNTAIN
OROVILLE
McCULLOCH
ROCK CREEK
CANADA
UNITED STATES

able to view the Spotted Lakes, a familiar landmark for early travellers. The Spotted Lakes, 5 miles west of Osoyoos, had been used for years by local Indians as a curative bath for rheumatism. Frequented by early settlers, their salts were used during World War I as a base for the manufacture of explosives.

The Kettle Valley line, a hustling, bustling operation was well patronized and was the quickest way to the world markets for the lumbering, farming and fruit-growing industries. And the cheapest. As the highways improved and motor transport trucks grew in popularity, they tended to weaken the stranglehold of the Kettle Valley Line. Thus, during the 1950's the struggle went on silently, almost unnoticed.

Nature and man combined to decide the struggle.

From Osoyoos the new Dewdney Trail wound over the breathtaking beauty of Anarchist Mountain through Bridesville and across the plunging Rock Creek gorge until it coasted down the Eastern rim to historic Rock Creek and on to Greenwood to sustain the heavy flow of tourist and freight transport traffic.

In the late 1950's and continued on into the 1960's the Department of Highways under "Flying Phil" Gaglardi put an end to the Kettle Valley line, a once romantic period.

Let me go back to where I began
If I can find the place.
It was far away and long ago,
And there was a kind of grace
I may not ever
Having grown old, recover.

This history of Canada 1905 – 1992, was developing hand in hand with "The Life of my Father" – a True Canadian.

In 1992 the Kettle Valley Railway and parts of the old Dewdney Trail were unfolding into a "Trail of Dreams", to join all the provinces across Canada to form a "Trans Canada Trail". This project is a legacy arising out of Canada's 125th Anniversary of Confederation and is directed by the Trans Canada Trail Foundation. This is a great beginning to a grand vision of developing railbeds into a multi-use recreational facility for the grandchildren and great-grandchildren of our early Canadian Pioneers who during the 1800's built those railbeds for all to travel along and to enjoy seeing all of Canada from coast to coast, not by travelling in wagon trains along dusty wagon trails, but in comfortable trains along the railroads. These old railbeds from the 1800's that will form the Trans Canada Trail will be for walking, cycling, horseback riding, cross country skiing, and snowmobiling. These trails will be exciting, interesting, well kept trails.

As a nation of trail builders, Canadians have done the impossible twice before. One of the greatest epics of railroad building started in 1880, and in 1945 the Trans-Canada Highway, from Victoria on Vancouver Island, British Columbia to Saint John's, Newfoundland. The government of British Columbia has acquired all the abandoned Kettle Valley Railway grade from Brookmere (near Hope) to Grand Forks in the west Kootenays. This ready built trail will be the backbone of the Trans Canada Trail in British Columbia. In Victoria where the Trans Canada Trail begins; At Selkirk Waters, a site for a shelter has been chosen where the old railway trestle
crosses the gorge. At this location, the city's hike and bike trails link up with the Capital Districts Galloping Goose Trail, then on to Nanaimo and to West Vancouver. The Dewdney Trail and the Kettle Valley Railway take over in Vancouver

and continue on to Hope–Princeton, to Penticton. Both the Kettle Valley Railway line of 1905 and the Dewdney Trail of 1860 met, and ran through Penticton.

In Penticton, a trail Pavilion is being constructed for names of all the people who are supporting the new "Trail of Dreams", and for names of the people who built Canada; who drove along the old Dewdney Trail in their Model T's, and the old timers who travelled the Kettle Valley Railway line.

In memory of my father:

His name: **Wm. Yawney 1905 – 1992**
The name of his store: **B-Y Store, Vernon**
1945 – 1965

Will be permanently inscribed and displayed on a special panel in the Penticton Trail Pavilion.

When you see my father's name – Wm. Yawney, on the Penticton Trail Pavilion wall, and if you close your eyes and listen: You may hear the whistle, and the rumbling sounds of a Kettle Valley train approaching Penticton – and the call "All Aboard". You may see a "Covered Wagon B-Y truck" creep through the city streets of Penticton, then slowly disappear into the early morning mist, with my father at the wheel, driving on to Princeton.

The Trans Canada Trail Pavilions will have a Canadian history to tell, for all travellers to remember.

My book – Title: ***Dear To My Heart*** – by Eugenia Irene, will be inscribed in the Penticton Pavilion below my father's name. My book talks about the life of my father; of Canada and of the Trails. From the sale of every book ***"Dear To My Heart"*** a percentage of the money will be donated to the "Trans Canada Trails" for as long as the book sells.

From Penticton the Trail of Dreams will run south to Rock Creek then east along the Boundary Country to Creston,

Cranbrook and on to the Elk Pass at the British Columbia – Alberta border.

Alberta became a province on September 1, 1905, the same year my father was born: April 30, 1905. His mother, Annie Yawney received a certificate from the Alberta government in honour of being one of the first pioneer settlers in the Canadian North; which at that time was the Northwest Territories. Today this certificate of honour is a treasure kept by my grandmother's only son, William's oldest daughter Anne Yawney-Chauvel. This certificate holds the memory of a brave, proud and strong great lady – a true Canadian Pioneer. My father's mother.

The concept of the Trans Canada Trail was conceived in 1992 and will be the longest trail in the world.

Kapeekwanapeekeepakeecheewonk

is the longest single word name in Canada; an Indian name given to a river rapids in Manitoba.

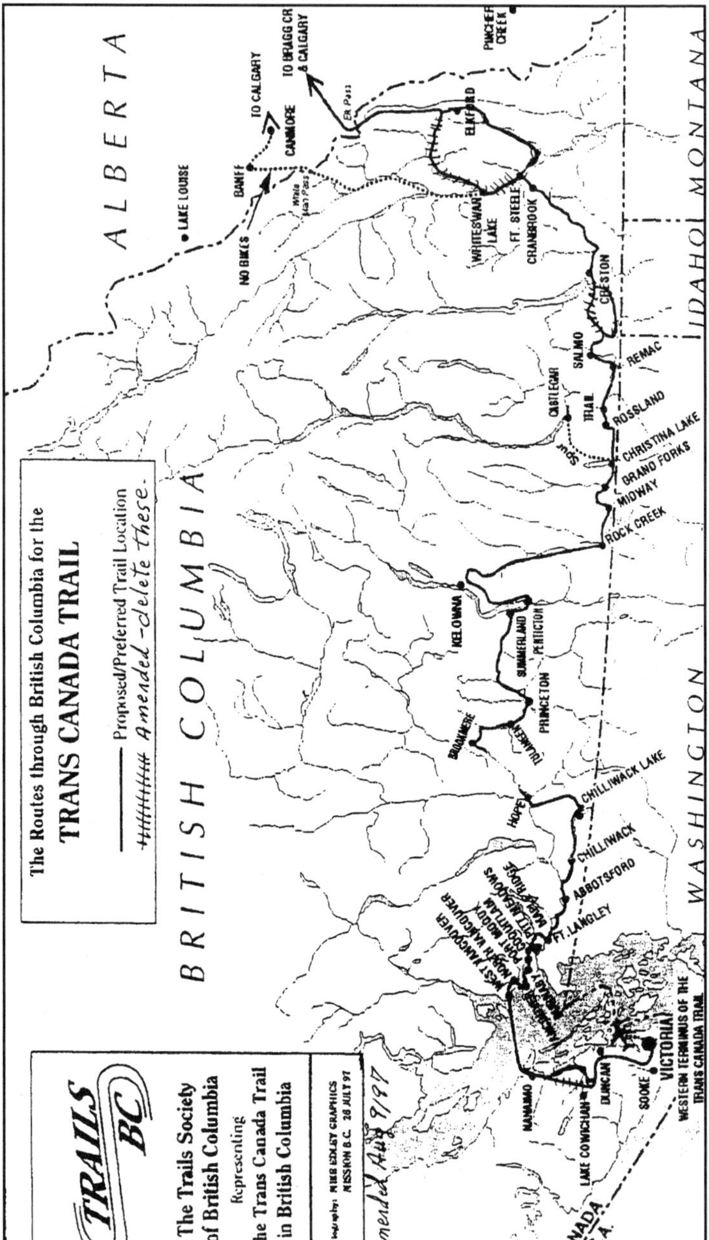
The Routes through British Columbia for the
TRANS CANADA TRAIL
Proposed/Preferred Trail Location
Amended - delete these.
TRAILS BC
The Trails Society of British Columbia
Representing
The Trans Canada Trail in British Columbia
Cartography: MIKE EDLEY GRAPHICS MISSION B.C. 28 JULY 97
Amended Aug 9/97
ALBERTA
BRITISH COLUMBIA
WASHINGTON
IDAHO
MONTANA
CANADA
U.S.A.
LAKE LOUISE
BANFF
CANMORE
TO CALGARY
TO BRAGG CR & CALGARY
NO BIKES
PINCHER CREEK
ELKFORD
WHITESWAN LAKE
FT. STEELE
CRANBROOK
CRESTON
SALMO
REMAC
CASTLEGAR
TRAIL
ROSSLAND
CHRISTINA LAKE
GRAND FORKS
MIDWAY
ROCK CREEK
KELOWNA
SUMMERLAND
PENTICTON
PRINCETON
BROOKMERE
TULAMEEN
HOPE
CHILLIWACK LAKE
CHILLIWACK
ABBOTSFORD
FT. LANGLEY
MAPLE RIDGE
PITT MEADOWS
COQUITLAM
PORT MOODY
NORTH VANCOUVER
WEST VANCOUVER
NANAIMO
DUNCAN
LAKE COWICHAN
SOOKE
VICTORIA
WESTERN TERMINUS OF THE TRANS CANADA TRAIL

Chapter 29

During the 1950's when my fathers farm – mid-way between Viking and Ranfurly in Alberta was sold; it was at that same time the Restaurant half of the B-Y Block closed. All the floors, walls, electrical and fixtures had to be ripped out and replaced with all new material. After the restaurant half of the B-Y Block was completely remodeled and all new, father rented that half to a flower shop. My mother said that half of the B-Y Block remained a flower shop to the day the building was sold in February 1966.

During those early years of the B-Y Store; my two grandmothers – my father's mother and my mother's mother together travelled by train through the Rocky Mountains to see how my father changed the life of his Katharine, of his sons and of his daughters from a life on a farm to a city merchants.

Travelling by train through the Rocky Mountains was quite an experience for the two old ladies who had not travelled outside of their rolling hills and the plains of Alberta since the day they first settled there with their spouses as young lady pioneers. Though they fretted and feuded in each other's company through the years, they made this journey together. The move my father made in 1943 by leaving his farm, a 14 year way of life and travelling through the Rocky Mountains in a boxcar to a place unknown, did not only change the lives of his Katharine, his sons and daughters, but the life of his mother. She came alone after that first brave experience of coming around the mountains in a train – she never would have seen the tunnels, the rivers, the beauty and the greatness of the Rocky Mountains; if her son, my father had not been on the other side. Every year without fail, with his Katharine at his side, my father travelled the trails, the highways and the by-ways to Alberta to visit his mother; though his Katharine rejected his mother, he loved his Katharine. And that special bond between a mother and her son remained within him all through his life. His last trip to Alberta was the summer of 1991, to see and to be, beside his mother's grave.

During the early 1960's my father purchased a new split-level, 4 bedroom family home on Mission Hill. The address was 3407 – 18th Avenue, Vernon. This new home was a few city blocks away from the 2 1/2 acres, on the Okanagan Avenue Road, where we had many picnics, where we danced and sang songs around a campfire. On the upper main floor of this home were three bedrooms, a bathroom, living room, dining room, large kitchen and a beautiful large sundeck that looked out over all of Vernon, Swan Lake, Silver Star ski mountain and all the orchards surrounding the city of Vernon, with the mountains beyond. Summer or winter, day or night; the scene from that sundeck was spectacular; more beautiful than from many homes anywhere that I have seen. At the lower level entrance was a patio with a picnic table. A door opened into a large family room with a piano, a television, a Marconi stereo system – record player, two comfortable chesterfields and chairs, a big picture window looking out onto the yard and a clothes closet. An antique old gramophone stood in the left corner of this room opposite the entrance door; a gramophone my father kept as a B-Y Store momentum for his first born grandson William. Through a door opening to the right were a bedroom, a bathroom, the laundry room, a pantry and a door leading out to the vehicle parking underneath the sundeck. In the yard were cherry, walnut, apricot trees, European cranberry, snowball and a special oak tree that never shed its leaves. There were flowers and more flowers, most of which were a variety of roses. And a vegetable garden my father cultivated. He brought in to that garden patch the best soil and natural organic fertilizers and the vegetables that grew there every year were spectacular and very tasty. Everyone who went there to visit, always filled their baskets full of garden vegetable goodies before they left.

With spring 1966 rapidly approaching, made my father feel ill and restless – more and more each day, he was missing his B-Y Store and decided to do something about it. ***He was taking his Katharine, the mother of his seven children on a world tour visiting all, or most, of all the countries on this world.*** May 1, 1966 at 9:00 a.m., my father, his Katharine and his son Patrick left their Apple Valley Home, in Vernon, on their way to Kamloops, then through the Fraser Canyon on to Vancouver with Patrick at the wheel. In Vancouver they visited friends and family for three days, and on May 4, 1966 at 5:00 p.m., my father and mother boarded a Cruise Ship, The Chusan. As a band played a farewell to all; their sons and their daughters stood on the platform deck waving white handkerchiefs and blowing kisses as the cruise ship with streamers flying down to us onto the deck slowly pulled away from the dock and was soon moving out towards the Lions Gate Bridge.

From a Stanley Park Lookout Point; speechless we all stood and watched as the cruise ship sailed away into the misty sunset and out of sight.

To see the world in a grain of sand
And a heaven in a wild flower
Hold infinity in the palm of your hand,
And eternity in an hour.

My father kept a complete diary of their world tour; of the mileage from island to island, and country to country. Of the sea temperatures and the land. The people they met and of where they went and of what they saw; out of which I will pinpoint the highlights only or my book will be a never ending story. This cruise my father and mother took was out of this world – or, I should say on this world. Most interesting and fantastic. During the 1960's very few people were able to travel, or travelled around the world. Holiday travelling during those early years was unique though as life went on into the 1990's, many more people today are travelling around the world. What is unique today, is travelling to the moon and into space. What a concept.

At 9:00 a.m. May 9th, 1966 the Chusan Cruise Ship docked in Honolulu Hawaii as a band played and girls in grass skirts sang and danced welcoming the ship into port. Every tourist received a lei of flowers as they departed the ship and onto a waiting tour bus. After seeing all of the sights of Hawaii and sampling and enjoying a variety of Hawaiian foods; at 12 o'clock midnight they sailed away on their way to Fiji.

May 14, 1966 as the cruise ship was crossing the International Date Line, there was no tomorrow. From Saturday, May 14th they skipped over into Monday, May 16th, 1966 and at 8:00 a.m., the ship docked in Suva, Fiji.

My father said that Fijian music and song is unbelievable, more beautiful than any vocal harmony singing he had every heard. The tour they took around the Fijian Island was to see botanical gardens, rice fields and they saw people working out on the fields with oxen, which took my father back 60 years in time when he, with his mother worked their homestead land with oxen. A Fijian band played on as their ship was leaving Suva and their calm beautiful romantic Island Fiji, at 5:00 p.m. May 16,1966 on its way to Auckland, New Zealand. Cruising through strong winds blowing over a very rough sea, they arrived in Auckland, New Zealand at 7:00 a.m., May 19th, 1966. In New Zealand they saw rolling hills of sheep and green pastures of cattle grazing. They spent a few hours at a natural hot spring, saw many wingless birds; the Kiwi, and too soon they were sailing away to Australia. My father loved New Zealand and her people and he said he would have loved to have stayed there a while longer.

The Cruise Ship Chusan docked in Sydney, Australia at 10:00 a.m., May 22, 1966. Sydney, Australia was the end of the line for the Chusan.

Lucy; a young lady from Vernon, British Columbia, had an ownership to a travel agency in Sydney, Australia. She met the Cruise Ship Chusan when she docked in Sydney and had my parents luggage transferred from the ship to a hotel awaiting the arrival of a P & O Orient Lines, the Ship Oronsay, on June 4, 1966. Their two weeks stay in Sydney, Australia was a relaxing and a welcoming break with no schedules. They were able to visit all of their daughter Ann's husband's family and friends and Lucy with her friends, took them to see many interesting places; to the Sydney Opera House. To operas, concerts, and to see a ballet. June 4, 1966, at 4:00 p.m., the P & O Orient Lines, the Oronsay left Sydney, Australia with my father and mother on board. The ship docked in Melbourne at 11:00 a.m., June 6, 1966. After a tour of the city and the country around Melbourne, the Oronsay left the port at 10:00 p.m. At midnight June 7, 1966, the ship docked in Adelaide, Australia, where the tourists again were off the ship to see the city, the parks and the beaches then sailed away at 2:00 p.m. The Oronsay cruised along the coast of Australia around Cape Leeuwin and docked in Fremantle at 7:00 a.m., June 10, 1966. This was a short stop; the holiday travellers went on a quick tour around Fremantle and Perth. They enjoyed the two cities, their beautiful parks and sights, boarded their ship Oronsay and sailed away at 1:00 p.m., the same day. Cruising the Indian Ocean. The Ship Oronsay was crossing the Equator at 3:00 p.m., June 16, 1966, where King Neptune got on board ship for a celebration and dance. Leaving the Indian Ocean, the Oronsay cruised the Bay of Bengal, past Coco Island and arrived in Colombo, Ceylon at 1:00 p.m., June 17, 1966. A tour bus took my parents to see many beautiful places, the slums, and to shop. June 18, 1966 the Oronsay cruised passed Cape Comorin, the southern tip of India into the Arabian Sea, cruising along the coast she docked in Bombay, India at 2:00 p.m., June 19, 1966. The Oronsay was cruising the Arabian Sea towards the Gulf of Aden on June 20, 1966: This day was my father and mother's 40th Wedding Anniversary. Telegrams from Canada arrived at the ship, and a special Anniversary Cake.

The Oronsay arrived in Aden, Saudi Arabia at 6:00 a.m., June 23, 1966. The city of Aden is built on a rocky shore as my father recorded and after a tour of the surrounding area of Aden they were on their way again sailing the Red Sea. At 1:00 a.m., June 25, 1966 the Oronsay docked in Suez, Egypt. Here; my father and mother departed from the ship Oronsay and boarded a coach line at 7:00 a.m. from Suez to Cairo. For two hundred miles they travelled the Arabian Desert and saw many herds of camel and donkeys roaming the desert. Around Cairo there were many oil refineries and glass factories. They were to see the Pyramids, and climbed a pyramid half way. They rode on camels, had lunch at the Cairo Hilton Hotel, then continued their coach line tour to Ismailiya and on to Port Said where they met their Cruise Ship the Oronsay. At 6:00 a.m., June 26, 1966 they were cruising the Mediterranean Sea and on June 28, 1966 at 8:00 a.m. the Oronsay docked in Piraeus, Greece. From Piraeus they toured by coach line to Athens. The history in and around Athens was mythological and well preserved since the 2nd century B.C. A most interesting place, like visiting another world. At 11:00 p.m. they sailed away leaving Greece with all her history behind. Cruising the Tyrrhenian Sea they arrived in Naples, Italy at 7:00 a.m., Thursday, June 30, 1966. Naples, surrounded by orchards and her mystical ruins of 1748. My father said good-bye to the Cruise Ship Oronsay and with his Katharine checked into the Paridiso Hotel. He arranged a European land tour with Aloschi Bros., Tour Lines, then together my father and mother sailed away in a little pea green boat to the Isle of Capri.

At 4:30 p.m., July 1, 1966 they were on their way to Rome where they checked into the Park Hotel. Their tour of

The Owl and the Pussycat went to sea
In a beautiful pea-green boat,
They took some honey, and plenty of money,
wrapped in a five pound note.

Rome started at 8:00 a.m., July 2, 1966, to the Vatican into St. Peter's Church, architecture of Michael Angelo. St. Peter's Church is 600 feet long and 400 feet wide and holds 60 thousand people. Then to St. Andrew's Basilica where Julius Caesar was stabbed to death at the age of 56 years. Recorded are many more places of interest, too numerous to mention. All the old history; the talent, the music, the art works and the architecture were amazing; very well preserved and very Italian. Italy is Italian. Sunday, July 3, 1966 at 8:00 a.m., they were on their way through famous hill-towns to Florence where they checked into The River Hotel. They toured all of Florence and Pisa where they saw the art collections of all of Michael Angelo's masterpieces. They took pictures of the Leaning Tower, saw small strip-farming and returned to Florence. July 5, 1966 they drive through Apennine Mountains to Bologna where they stopped for lunch. Along their way they saw oxen in yoke, and the village where Marconi was born, then crossed a river to Venice where they checked into the Cappelli Hotel. July 6, 1966, at 8:00 a.m. they were on their way to Verona where they stopped to see the home of Romeo and Juliet and to have lunch in the Barchetta Hotel. They continued on to Milan, an industrial city, arriving at 4:00 p.m. From Milan, Friday, July 8, 1966 they continued on into Switzerland, and stopped for lunch in a hotel in Berne where they met a school principal from Winnipeg, Manitoba and a teacher from Vernon, British Columbia who was teaching there in Berne. Their trip continued on through a 9 .5 mile tunnel, and they arrived in Lucern, Switzerland at 7:00 p.m. the same day. Saturday, July 9, 1966 after sightseeing and shopping around Lucern, they continued on to Zurich along beautiful valleys and over mountains. They saw many farms on top of the mountains and many jersey colour cattle.

"Very nice", father said. Zurich, the capital city of Switzerland, is a beautiful industrial city. July 10, 1966, after a tour of the city, they were on their way through the Black Forest, West Germany. Stopped for lunch in a hotel in Pforzheim, a cattle country in very good condition, and arrived in Heidelberg at 7:00 p.m. Heidelberg Oh! Heidelberg; at 9:00 p.m. they went to see a concert at the Heidelberg University where the musical "Student Prince" was filmed in 1954.

July 12, 1966 at 7:00 a.m., they say good-bye to Heidelberg and drive along Heidelberg country to Main where

they board a Rhine River Steamer, and for forty miles they cruise the Rhine River, have lunch on board then rejoin their coach, drive to Bonn and arrive in Cologne at 5:00 p.m., where they checked into the Hotel Berlin. After dinner and a tour of the city, at 9:00 p.m. they go to see a University concert.

Wednesday, July 13, 1966 at 11:00 a.m., they arrive in Brussels, capital of Belgium, a beautiful city. The Exhibition Towers of the World Fair of 1958 still stand; father recorded. They drive on to Mons to see the 1914 – 1918 World War I memorial. Cross the border into France, have lunch in Reims and arrive in Paris at 7:00 p.m. After a relaxing dinner they went to Bois De Boulogne.

July 14, 1966, the whole day was Paris. To be in Paris and to see all of Paris. My father recorded all of Paris in French, so I'll just say: He was in Paris with his Katharine and they saw it all. That evening they went to the Theatre of Music – Le Moulin Rouge to see the Opera Illuminis.

I Love Paris

I love Paris in the springtime,
I love Paris in the fall,
I love Paris in the winter, when it drizzles,
I love Paris in the summer, when it sizzles,
I love Paris ev'ry moment,
Ev'ry moment of the year.
I love Paris,
Why, oh, why do I love Paris?
Because my love is near.

July 15, 1966 at 6:00 a.m., they leave the excitement of Paris and are on their way to Beauvais, to Abbeville and on to Calais where they stop for lunch. Here in Calais they embark a steamer to cross the English Channel – The Strait of Dover, to Dover, England. In Dover they hop a train to London and hail a taxi to the Imperial Hotel where they stay for the next five days to see all of England. And they did. They rode the subways, walked in the parks, and went on a sightseeing tour along the River Thames. To the Tower of London, Trafalgar Square. To Buckingham Palace to see the changing of the guard. Westminster Abby and Cathedral, the Big Ben, The House of Parliament. Lambeth Palace, home of the Archbishop of Canterbury. They had tea in a 400 year old Grimm's Kitchen, and took pictures of grass thatched roof houses in Wimbledon. A bus tour took them through, around and over the London Hills and to see the London University. In his diary my father recorded many, many more places of beauty, and of history, of which I will not keep writing about. - *"One must see it to believe it"*. England is history – not to forget the English Country Gardens.

July 20, 1966 at 11:55 a.m., my father arranged a flight with Air Canada, and they arrived in Toronto, Ontario. He did not record the time. Their son Henry met their flight then transported them with their luggage to his home in Grimsby, Ontario, where they stayed for the next two weeks in care of his five sons, three daughters and his home, while Henry with his Joan went off to New York City, New York, on a two week holiday and shopping.

During the first week in August 1966, when my father and mother returned to their Apple Valley Home in Vernon, my father got right to work to remodel and rebuild an old pioneer shack on ten acres of commercial property, in the city of Vernon, he had purchased during the years he was in the B-Y Store business. He purchased this ten acre property: Address – R.R. #4 – 5798 Okanagan Landing Road, Vernon, B.C. at $8,000.00 per five acre parcel. A total of $16,000.00 for the ten acres. A quote given to me by my mother, the year 1996. I was not shown the records, therefore cannot confirm the purchase price. This ten acres was an alfalfa field with a

Father starting to rebuild shack on 10 acres.

The cedar bungalow built by Patrick and his father.

Connie, Glenn, Patrick and the horse. Father with grandson William and Craig.

Father on 10 acres under an apple tree.

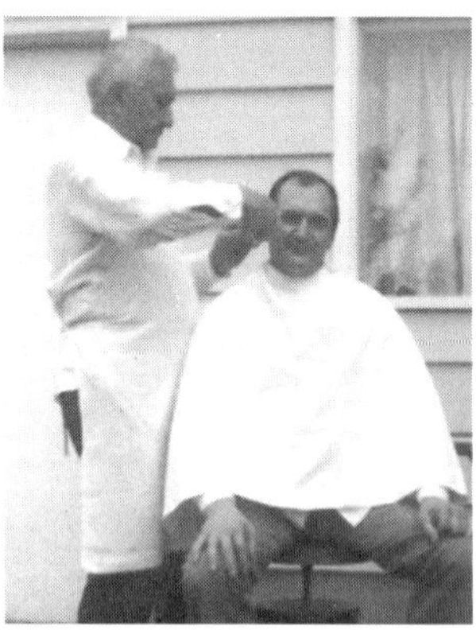

Henry

Father the barber

Edward

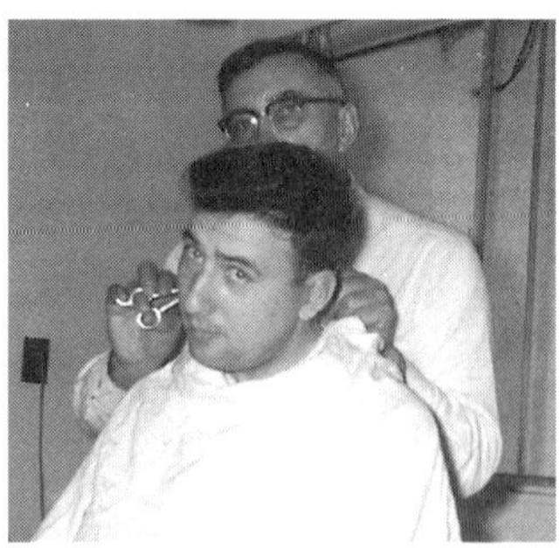

pioneer shack surrounded by apple trees approximately two-thirds back of Okanagan Landing Road. The back third, to the Vernon Creek was a pasture for horses, with a marsh to the right along the creek. This marsh was a home to many Canada Geese who nested there every spring. Patrick was just married, and with his bride Connie, their home was a suite in a new apartment building on 30th Avenue west of downtown Vernon. My father challenged Patrick to work with him. To tear down the old shack and to build a new home for Patrick and his bride Connie. Patrick was not keen on living on acreage or tearing down and rebuilding an old shack; though he listened to his father and together they got to work. My mother was there every day with their lunch and snacks in the afternoon. Patrick's bride Connie would have nothing to do with it, she sat in their car reading a book every day, while Patrick with his father worked. After the cedar bungalow was built, beautiful and with a carport; my father signed five acres of that ten acre parcel to Patrick to start making monthly payments on. Patrick with his Connie then left their apartment way of life and moved into that new cedar bungalow; built for two. After three sons came into their lives; Craig, Glenn and Dean; a split level with more rooms and an upstairs was added onto their cedar bungalow. It was then that my father signed the remaining five acres he held, to Patrick.

This 10 acres was near the Vernon airport, between the city of Vernon and the Okanagan Lake. There; it was quiet and serene with the sounds of a variety of birds chirping through the trees. You could hear the rippling sounds of the Vernon Creek from the screened in deck facing the back yard and from there you could watch the horses playfully romp about. Patrick kept this land very well groomed, like a park.

This was a paradise.

1965 – 1995, — 30 years later that property assessed value was well over a million dollars. I will not reveal the exact sales price of this ten acre property, as Patrick wishes to keep that amount confidential, and I honour his wish. Why I did mention this 10 acre parcel is for young people of today and tomorrow to realize the foresight and the wisdom of my father. He knew how to survive through the hardest times: The pioneer years, the depression years, through two world wars when many people had to live on food ration tokens alone to survive. My father chose to raise his young family on a farm; a land of plenty; though he was a business man at heart.

The road through life for a man is rocky and rough, and more so when he has to pacify a demanding, jealous, complaining spouse which most men seem to be laden with. Through their journey of life – some men, the strong ones make it. Many do not. My father's will was to journey through life, one step at a time. To climb a mountain one must start at the bottom to reach the top. Those who chose to start at the top, fell!! And they fell hard. The one son my father challenged, made it to the top.

My father kept a journal of his lifelong works. On one page from the beginning, of this journal, was the date 1924, from when he was 19 years old. This would be of his savings from his work at the paper mill in Washugal, Washington and perhaps his savings from the first beaver pelt he ever sold. That is all that was left of his journal. The date and the cover, the inside pages were cut out and destroyed not to be known of, or recognized. The ingenuity and the history this man, my father, recorded in his journal in gold (I say in gold – they were words of gold) are gone forever. He always said "knowledge is golden". His journal was a book of knowledge to be shared and to be treasured and perhaps be kept in a museum as a reference of his foresight since the days of his youth. – **DESTROYED** – Why? He did not destroy it.

We must remind the world of the great men of our time and that is what I have done in this book "Dear To My Heart". To write only of our past and present Premiers and Prime Ministers – great men – is mind-boggling. What they have done, allowed, are allowing and doing to our Canada and her people is disheartening and sad.

For example in reference to NAFTA and MAI. The Ethyl Corporation, a U.S. transnational, has filed a $350 million lawsuit against the Canadian government. This has been done under a provision of NAFTA (North American Free Trade Agreement) that grants corporations private legal standing and the ability to sue governments. The crime of the Canadian government was that parliament banned the importation of MMT, a neurotoxin additive to gasoline. The Ethyl Corporation, which is the sole manufacturer of MMT, claims this is tantamount to an expropriation of their property, and they must be compensated.

NAFTA is an agreement among Canada, the U.S.A. and Mexico. Brian Mulroney, a Tory, set the North American Free Trade Agreement in motion in 1988, between Canada and the U.S.A. Jean Chretien, a Liberal, added Mexico to NAFTA in 1994.

The Multilateral Agreement on Investment (MAI) extends provisions of NAFTA to the 29 countries of the Organization for Economic Co-operation and Development (OECD). It is my opinion that MAI should not be passed and that NAFTA should be recalled. Both are dangerous to Canadian sovereignty and interests.

1995, Brian Mulroney, a Conservative Tory, who resigned as Canada's Prime Minister in 1993 filed a $50 million lawsuit against the Canadian government, plus costs of another $2 million. He believed he too had the right to sue the Canadian government. Brian Mulroney launched the unprecedented lawsuit after a 1995 letter from the federal justice department to Swiss authorities alleged Mulroney received millions of dollars in kickbacks in Air Canada's 1988 purchase of Airbus jets and two other contracts. This money he put into a Swiss bank under a fictitious name. To prove that he was, and is a true Tory – Tory which means a Robber – A "Robin Hood" who robbed all Canadian people for the likes of himself. In reference to the Airbus investigation, the Government of Canada and the R.C.M.P. had reached conclusions that Brian Mulroney had engaged in criminal activity. The Airbus investigation was a highly confidential state-to-state communication intended to be received by Swiss authorities. To this day, January 6, 1997, it is not known how the newspaper that first reported the matter obtained a copy of a non-governmental summary of the document. The Government of Canada did not foresee that the Request for Assistance would become public. Since it did, when public money is involved, the Government of Canada has reviewed its procedure to ensure that the risk of this happening again is minimized before it is to become public knowledge. To know Brian Mulroney; you must read: **"ON THE TAKE - Crime Corruption and Greed in the Mulroney Years"**

by Stevie Cameron, Canada's leading political journalist.

1998, Jean Charest, a loyal Canadian Conservative, made a switch from being Canada's Conservative leader in Ottawa, to being a Liberal in Quebec. To win the hearts of all the Canadian people from coast to coast to coast, and he would if he would make another switch and become the leader of a new party – The **CLOND** - (Conservative Liberal or New Democrats). In our Canadian parliament; all that our - great men - our Prime Ministers and all our politicians do, is BLEAT and BICKER.

Are they the dignified, the respectful? Are they a perfect example to follow and to be proud of for all Canadians? — Why students are fighting and killing each other. Is this not what the United States is doing and Canada is following suit? Bill Clinton of the United States does not like Saddam Hussein of Iraq; he sends all U.S.A. modern helicopters, bombers, ships and troops to the Persian Gulf to wipe Hussein out of power; to kill him. And Jean Chretien of Canada is right there willing to help Clinton do that. Both U.S.A.'s Clinton and Canada's Chretien make it quite clear that Fidel Castro of Cuba should be wiped out. Why? Because they do not like him.

Chretien allowed a conglomeration into our country, and many of whom are criminals. These foreigners are behaving in the same manner as Clinton and Chretien. They have no respect for our Canadian way of life: Our police force or our system. And they kill the people who they do not like. No foreigner should be allowed in our Canadian Parliament before they have abided by our Canadian laws and rules, for at least two generations.

Chapter 30

My Thoughts of the 1990's

Give. Give. Give. Give. Give. — Is going much too far. To pick up the morning paper, turn on the radio, the television, answer the telephone ring or the knock on your door; and it's give, give, give.

To receive something for nothing. – Is worth nothing.

Money, property, the generosity of a giver is more often than not squandered. – Not appreciated by most receivers but gives the receiver strength to become more inclined to resist and to oppose assistance to whomever may really be in need. Youth of the 1990's are taking advantage of our governments "Welfare Fund", "The Law" and of the "Human Rights". To give, to support "Amnesty International"; is to pardon criminals, and disgraceful behaviour.

LIFE

One must live a life to understand life.

How can a child with no experience in living a full life be allowed to demand a "Right".

To undertake, to experience, and to live an independent, responsible, self reliant way of life is to know life, and it takes a lifetime of experiences. This story of the Little Red Hen explains…

"The Little Red Hen"

"Once upon a time a little red hen lived with her chicks in a small cottage. She worked hard to keep her family well fed. In the evenings, she sang while she worked.

One Day when the little red hen was out walking with her friends, the goose, the cat, and the pig, she found a few grains of wheat.

"Who will plant this wheat?" she asked her friends, the goose, the cat, and the pig.

"Not I," said the goose.

"Not I," said the cat.

"Not I," said the pig.

"Then I'll do it myself," said the little red hen. And she did.

One morning the little red hen saw that the green wheat had sprouted.

"Oh, come and see the green wheat growing!" she called to her chicks.

All summer the wheat grew taller and taller. It turned from green to gold, and at last it was time for the wheat to be harvested.

"Who will harvest this wheat?" she asked her friends, the goose, the cat, and the pig.

"Not I," said the goose.

"Not I," said the cat.

"Not I," said the pig.

"Then I'll do it myself," said the little red hen. And she did.

At last the wheat was all cut down and it was time for it to be threshed.

"Who will thresh this wheat?" she asked her friends, the goose, the cat, and the pig.

"Not I," said the goose.

"Not I," said the cat.

"Not I," said the pig.

"Then I'll do it myself," said the little red hen. And she did.

At last the wheat was threshed, and the little red hen poured the golden grains into a large sack, ready to take to the mill to be ground into flour.

The next morning the little red hen asked her friends, the goose, the cat, and the pig: "Who will take this wheat to the mill to be ground into flour?"

"Not I," said the goose.

"Not I," said the cat.

"Not I," said the pig.

"Then I'll do it myself," said the little red hen. And she did.

The next day the little red hen asked her friends, the goose, the cat, and the pig: "Who will bake this flour into a lovely loaf of bread?"

"Not I," said the goose.

"Not I," said the cat.

"Not I," said the pig.

"Then I'll do it myself," said the little red hen. And she did.

A government paid for taxi ride to the Welfare office and to the food bank gets it all; for the goose, the cat and the pig.

At last the bread was baked and the little red hen called to her friends, the goose, the cat, and the pig: "Who will eat this lovely loaf of bread?"

"I will!" said the goose.

"I will!" said the cat.

"I will!" said the pig.

"Oh, no, you won't!" said the little red hen.

"I found the wheat and I planted it. I watched the wheat grow, and when it was time I harvested it and threshed it and took it to the mill to be ground into flour, and at last I've baked this lovely loaf of bread."

"Now," said the little red hen, "I'm going to eat it myself."

And she did!"

From the disasters of the world wars and the depression years; Canada has leaped right straight into another disaster by allowing "Human Rights", "Children's Rights" – (children to become pregnant, to break into other peoples homes, to steal cars of which are not their own. And to kill. All for which they are not punished, because they are children.) Is this Right???? The "Welfare Fund" has taken away the dignity, integrity, the willpower and the creativity out of Canadian young human beings to learn to be able to stand on their own and to be responsible. To see all the protests and parades throughout all of Canada is shameful, degrading and unbelievable. For the very most important position held in our country; the people of the Provinces and of Canada; vote in a leader; then protest against everything he tries to do. We must know of his honesty, honour and compassion before we vote him into power. 1996; Jean Charest said, and I quote: "I want to leave for my children, the Canada my parents left for me." Dear John: — I do hope you can do just that. If our ancestors could come back to life, they would die again from disbelief, seeing what their past labours have today become.

Here are words from a book title: Ten Lost Years 1929 – 1939 by Barry Broadfoot – A woman's words: "What it was like out there in the Ontario bush without more than a couple of dollars from one months end to another. She tried to tell them because she felt they should know. That everyone should know – EVERYONE – her little brother was going to school wearing girls clothes, because the only clothes he had to wear were the hand-me-downs from his sisters and from her." During those early years, the people of Canada, like the woman in the Ontario bush, and her little brother who grew up wearing his sisters hand-me-down-dresses, raised their children to be independent, reliable and responsible citizens. Today: These are the people: The only people who are supporting all the takers. Before long, there will be no givers left to give, because the takers are raising their children to be takers too.

During the 1950 – 1960's, my father's first born grandson, (my son) was growing up without a television to watch or having toys to play with. And there were other children like him. My son got one second-hand story book "Farm Stories", a 1946 Teggren-Jackson book when he was three years old and was able to read. For Christmas or for his birthday he got clothes 2 – 4 sizes too big to have plenty of room to grow into. And when he was growing out of those clothes and had worn holes on the seat of his pants, his knees and on his elbows; he got iron-on-patches, ironed on because there was no time to sit, to sew on patches. Work was never done: On the job, in the home, by growing a garden to live on and preserving, fruits, vegetables and meats for the winter months. To be self reliant was instilled into every

William in snow suit
2 - 4 sizes too big.

young human being; to be a True Canadian. Since 1969 those rules have come to an end.

"It is not what your Country can do for you –
It is what you can do for your Country."

How do you interpret these words of John F. Kennedy?

All the Human Rights our hard working Canadians since the Pioneer years were looking forward to; in the 1990's are being misrepresented and misinterpreted. The Tommy Douglas's, the Mickey Rooney's, the Apple Annies have tried to put a true picture into words and into focus which our chosen government leaders have overlooked leaving the generations of the woman from the Ontario bush, of her little brother who wore hand-me-down girls clothes, my son and the young responsible generations of the 1950 – 1960's, to carry the load. **"R I G H T S"** – think again.

Many of our Canadian homeless are the honest, trusting hardworking citizens of the past. Some are business men and women, some are farmers and some are professional men and women who have laboured most of their lives, paid their taxes and supported their families. Because of a Welfare fund and Woman's Rights that brought on divorce; together with our corrupt legal system with no justice; put these elderly people, left with nothing, out in the cold. And; these are the people who do not ask for any kind of help, because they always gave and do not know how to ask. These are the Mickey Rooney's and the Apple Annies who truly do need, deserve and have earned government help – They are the forgotten ones.

The 1990's wage earner…From an average income

- 30% is paid into Income Tax
- 30% is mortgage interest
- 10% is for other government fees and interests.

A total of … 70% out of every worthless looney is paid out of their wages before they can start supporting their own families. 20¢ - 30¢ from every dollar they earn is all that is left for them; the wage earner. Are they to give to support the takers?

Our Parliament body keeps adding more and more taxes to all goods required for life and for living, which puts a demand on all wages to become higher and higher to meet the workers needs. These taxes are the PPT – Property Purchase Tax; GST – Goods and Services Tax, PST – Provincial Sales Tax, plus all other taxes which do not encourage our young people to want to work, and they become wards of the government.

Some are put into foster homes, and some are left on city streets.

The youth who are on city streets and on drugs; should be scooped up into crates and hauled out to some island with no way out; to fend for themselves. The ones who would live; would be brought back to live like human beings: The ones who would not live, would die of fresh air, sunshine and rain; Not from an overdose of drugs and by draining government funds. Human beings should behave like human beings and when they will be left alone, to fend for themselves; they will soon learn to stand on their own.

Leave them alone and they'll come home and bring their tails behind them.

To be cruel – is to be kind.

All people who are on drugs all over the country should not be allowed on city streets, doorways and back alleys injecting themselves with drugs. They are a disgrace to our citizens and cities. No shelters should be built in cities for drug addicts; and the drug pushers, should be put in cells and injected with the drugs they sell; slowly and every day until that drug they sell to others – kills them a slow death. Today; crime and criminals are overpowering the world. Every criminal should be exposed in every newspaper and on every television station so that the public knows and sees and is made aware of who they are. They must be punished however young. If Parliament and our government would abolish all welfare, and free needles to drug addicts. Cease to sue tobacco companies, or include the Chemical Companies, Pulp Mills, Oil Refineries, Automobile Manufacturers, and there are more, sue them all for polluting our world. To set-up a subsidy towards a property of choice; for the business minded young folk, and for young country minded folk to help them build their goal in life: That would be the honourable and compassionate thing to do. Not by destroying young lives with Welfare, drugs and free needles.

Canada.

One hundred and fifty years ago the 1800's – is there a difference?

The 1800's – white man gave liquor, disease and bibles to the native Canadians. Today the 1990's, men and women give welfare, drugs and free needles to Canadians. – ***A thought.*** ***? ? ? G I V E ? ? ?***

In the early 1800's foreigners, many with young families from Europe and China arrived by ship under horrid conditions to begin a new life in Canada. A new life of extreme hardships. These courageous pioneers should be remembered and honoured for the education of all Canadians. In our schools to re-live these courageous people would build self-esteem and leadership qualities and set a foundation for youth achievement. This journey into the past would generate an interest and appreciation of our pioneers unique past; those brave people who built and paved the way, for Canadians of today. Building our roads and railroads in the 1800's through a most hazardous mountain terrain in British Columbia, many pioneers, most of whom were Chinese died because the work was very hard and conditions were very poor. When the sensational rumors of our rich gold fields reached California, tens of thousands of opportunists trekked northward to flood our country. Young pioneer Canadians were dying from hard work building our roads and railroads, while American opportunists, using our roads and railroads came for the taking. Times have not changed since the 1800's, opportunists from the United States continue to flock into Canada and to help themselves to all Canadians natural resources: Our minerals, water, fish, and our land; as if it all is theirs for the taking.

Springtime 1973; my father, mother and I were at the Vancouver International Airport waiting to board a flight to Las Vegas, Nevada. At the desk of the customs officer; my father was asked – "Where are you travelling to?" My father replied – "To Las Vegas! With my wife and daughter to watch my grandson play college football." The officer then asked – "Who did you vote for during your last election?" My father was taken aback by that question; and being the wise business man he always was and never discussed politics or religion in business said "Dave Barrett got the majority votes: He is the Premier of British Columbia." That U.S.A., S.O.B. customs officer did not like my fathers reply and did not allow my father to board our flight. He said that my mother and I may go, NOT my father. The three of us went back to the main waiting room from where I telephoned my son's football coach, Mike Caviness, who had our reservations made for our stay in Las Vegas with the football club; and a car was there waiting at the airport to pick us up. Coach Mike Caviness, instructed us to drive to Seattle and to take a flight to Las Vegas from there. My father was an honourable man, never a threat to anyone and never, absolutely not, one to sneak around. Taken aback, surprised, with feelings hurt he refused to go if he could not board the flight in Vancouver: with his suitcase in his hand and his binoculars and camera hung over his shoulders; we returned to my home in Coquitlam. My father's eyesight was rapidly failing him and the reason he desperately wanted to see his grandson play his college football. The following morning my father insisted that my mother and I together must go, not to disappoint his grandson by us not being there to watch his team the "Ogden Utah Wildcats" play Las Vegas that afternoon. My mother and I left my home early and after I parked my car in the airport parking we hurried to the U.S.A. customs desk and there stood that same U.S.A., S.O.B. and this early morning, he chose not to allow my mother nor me, to board our flight to Las Vegas. That was a cruel act. After all that Canada has allowed the Americans since the pioneer years, and through the war years; where Americans using our land – NO charge left all their garbage and pollution for Canadians to clean up.

1996 – 24 years later; our Canadian government is once again about to sell to the U.S.A., all that we cherish the most; our Canadian farm lands, without so much as a squeak in the public arena. With the Canadian Looney worth half of a United States dollar; this is not a sale of Canadian lands; this is a give-a-way.

When we Canadians travel along our Trans-Canada Highways, that our ancestors built for us, to see our land; all that we will see, will be American farms on Canadian soil. The Americans, we Canadians refer to as our neighbours are takers, manipulators, dictators, and are trying to monopolize the whole world while our great government politicians (what a lot of numbskulls) for a personal power and glory are allowing them to do so. And; we all,

We put them there!

All abandoned Canadian farm lands should be acquired by our government for a ranch, a lodge, a commune to be built by, and for, all Canadian homeless people; to grow their own gardens, to learn to milk cows without using milking machines. To experience the joys of seeing new born calves in the spring season of every year, tiny pink baby pigs, little lambs and kids. To learn to set a clucking hen on 10 – 12 eggs then after 3 weeks watch the baby chicks as they hatch then climb out from under the mother hen, and with her help try to leap up onto her back to bask in the warm springtime sun. To ride horseback through woods and open country. To stop and to have a picnic beside a babbling brook and perhaps catch a trout before returning home. There is so much life, and of life to learn about on a farm instead of standing around on city streets and city back alleys injecting oneself with deadly drugs. To keep our land for our homeless, to live together, to build and to share; our government would not approve of because that would be Communism; and Communism, we Canadians are taught and told to be afraid of and not to accept. Instead; our politicians choose to hand out free needles to our youth, to keep them lost and cold on our city streets. I believe that a variety in government is as necessary as is a variety in our diet and on our dinner plates.

Not the politicians; but the scientists, teachers, professors: - these are the professionals who should, and are able to, and who have the knowledge to, warn the public and all students of the dangers of all drugs, chemicals and pesticides, all of which are in all of our foods, water, coffee and in tobacco and all are a grave hazard and equally dangerous to all our lives. If our politicians would only learn to listen to the people. Not to dictate but to listen and to hear; as the wise old owl:

who lived in an oak;
The more he saw
the less he spoke;
The less he spoke
the more he heard.
Why can't we all
be like that bird?

To listen to the cries of the people, the animals, the mammals and the birds who are dying, and not from smoking cigarettes. Every child, every individual is unique and must be listened to, and to have a "right to choose". To choose a way of life and to know that they and they alone must be responsible for the choices they make. The 1990's – science knowledge, modern medicine, ultra sound, available abortion clinics – these should be personal and private choices. As should euthanasia be a personal and a private choice. Institutions across Canada are loaded with abandoned disabled children. And that is cruel. Parents should either keep them or abort them before birth. Rights to choose are personal. If those choices are interfered with; the life of that individual is destroyed for a lifetime. However young, and however old, if an individual needs advice or guidance, he will ask for it. What a person really needs is a good listener. Young Canadians have much more knowledge, ability, compassion than most adults or politicians do, who are dictating to them and driving them to smoke, drink and to take drugs. Politicians should show our young Canadians kindness, respect, appreciation, trust and a freedom of choice. They do need a good listener; not a dictator and a free needle.

Working parents are too busy with their lives, and politicians are too busy living.

The 1990's – O Canada
Let us keep our land.
Our own beloved native land!

Chapter 31

June 1965 my father's youngest son Marshall James graduated from the University of British Columbia, and together with my father they celebrated my father's 60th birthday and Marshall James graduation over a gourmet dinner served for all the family, their spouses and their friends at the new Blue Boy Motor Inn Hotel, on Marine Drive and Fraser, on the outskirts of the city of Vancouver, overlooking Lulu Island. Leaving the University of British Columbia, Marshall James was travelling on to a far away land to work towards a Ph.D., together with his lady friend, Vera Jessie – who named herself Jo. Their chosen school of hard knocks was the University of Moscow, in the U.S.S.R. The Union of Soviet Socialist Republics. The University of Moscow did not accept young students, who were not married to live together in their university dormitory: With their hearts set to continue their life of learning in Moscow; a quick wedding was arranged, and they were married in a little granite rock built church in Vernon, British Columbia near the City Hall, with Jo's granny McLeod playing their wedding march on the church organ. Jo was an only child raised by her mother, Jean Lear, and Marshall James being the youngest son of William and Katharine Yawney; both families were pleased and relieved to know that Marshall and Jo, during their years away from their home in Canada, will have each other to rely on. Marshall was never a person to brag, and through the years his description of Ph.D., he said meant "Piled Higher and Deeper"; which in reality is the holder of the highest University degree in any faculty. Since the wedding march was played by Jo's granny McLeod in that little rock church; Marshall and Jo's marriage has been as solid as a rock. — Was it their chosen church; their love for each other; or is it the knowledge and their life they had away from their home land that made a solid foundation for their life together during these years and times of uncertainty. Alone, they hold the reason; that secret belongs to them.

With Expo '67 on in Montreal, and Marshall and Jo in Moscow; my father with his Katharine were on the road again. Always my father's concern for his family, and his determination to view his young sons way of life in a foreign land, they left Vernon by motor car on August 6, 1967, driving to Vancouver where they boarded a jet aircraft on August 8, 1967 and arrived in Toronto at 4:30 p.m. After a week's visit in Toronto, Hamilton, Niagara Falls, Kitchener and a visit to the Stratford Theatre, they boarded an aircraft at 9:30 a.m. August 14, 1967 and arrived in Ottawa, Ontario at 10:25 a.m. where they went site seeing for the day, then boarded an aircraft at 5:35 p.m. arriving in Montreal, Quebec at 6:10 p.m. From the airport they took a taxi to the Alexander Pushkin cruise ship where they were directed to their cabin for their Atlantic Ocean voyage to Europe, and on to the U.S.S.R. August 15, 1967 they took a taxi from the cruise chip to Expo '67, to see "Man and his world, and man in space". First U.S.S.R. man in space; and they went to the U.S.S.R. and the Algerian Pavilions before returning to the ship. August 16, 1967 they were back at Expo '67 to see the Canadian Pavilion, the Canadian Native Indian Pavilion, the French and the American Pavilions and they took rides on the Electric Train. August 17, 1967 was their last day at Expo '67, and a full day it was visiting pavilions: the Australian, India's, Mexico's, Burma, Egypt, Ontario, Quebec, Ceylon, Germany and Czechoslovakia.

August 18, 1967 at 11:00 a.m. the Alexander Pushkin cruise ship sailed away with my father and mother on board leaving Expo '67 and Montreal, Quebec behind them. Cruising along the St. Lawrence River; my father recorded in his travel journal that the countryside on the banks of the St. Lawrence River was beautiful. Thoroughpaced; on into the Gulf of St. Lawrence. August 19, 1967 at daybreak, they found themselves cruising along on the Atlantic Ocean. Father recorded in his journal, that the meals on the Alexander Pushkin ship were very good; enjoyable. August 20, 1967 they were sailing along through a heavy thick fog with a strong wind blowing; on August 21st the day was sunny and warm and on August 22nd again they sailed through heavy fog with cool temperatures. August 23rd sunny and cloudy, and on August 24th misty and warm. August 25, 1967 they were sailing along slowly with the weather calm and warm. August 26, 1967 the Alexander Pushkin cruise ship docked in Tilbury, near London, England at 7:00 a.m. From the cruise ship my father and mother took an Electric Train into London where they met their interpreter, and from there they took a taxi to the airport terminal; and they waited. Their jet was two and a half hours late arriving. They boarded a Yugoslavia jet at 6:30 p.m. arriving in Zagreb, Yugoslavia at 10:00 p.m. After a half hour stay in Zagreb, they boarded a flight at 10:30 p.m. arriving in Beograd at 11:15 p.m., their interpreter with them. From the air terminal they took a taxi to the Hotel Slavia. My father recorded in his journal; "The Hotel Slavia is a government managed hotel; it is very clean, and the price for two people per day is $7.50." They stayed in Beograd for four days, went on tours and saw old forts from 12 centuries. August 30, 1967, my father and mother boarded a bus in Beograd at 9:30 a.m. and arrived in Sarajevo, Yugoslavia at 6:00 p.m. They found a very clean private place to stay in and on August 31st they went site seeing. September 1, 1967 they went to a bazaar and a museum – Gavrilo Princip. September 2nd they boarded a bus in Sarajevo at 11:00 a.m. for their journey over the mountains to Dubrovnik, a health resort on the shores of the Adriatic Sea, arriving there at 6:00 p.m. In Dubrovnik they check into a motel, The Majkovica Marko Sipcine No. 4. September 3, 1967, by street car, they went site seeing through the city. September 4th during the morning hours, they mailed a letter to each; their son Patrick in Vernon, British Columbia, Canada; and to their daughter Joyce who arranged this health resort excursion for my father and mother from her private personal Travel Agency. The afternoon hours were spent at the beach. September 5th after an early breakfast at the Majkovica Marko a thunderstorm with heavy rain stormed in and stayed all day, keeping them from going to the beach. September 6, 1967 was a nice sunny day relaxing at the beach Lapad the whole day. September 7th was a hot, sunny day relaxing at the beach and taking pictures. September 8th was a fair day relaxing at the beach Lapad, and September 9th was another fair day they enjoyed at the beach. Sunday, September 10, 1967 was a cloudy day with thunder, lighting and heavy rain. My father with his Katharine went on a tour to Old Dubrovnik, an old historical fort city inside the fort. They ate good meals and listened to music in a modernized hotel before returning to their motel in Dubrovnik. Monday, September 11th at 7:30 a.m. they boarded a bus to Budva. My father recorded in his journal "They enjoyed a beautiful scenic drive along the coast", "that Budva has an old historic city in a fort and five beautiful modern hotels". The bus tour to Budva and return to Dubrovnik was $67.20 for two, and their dinner in a modern hotel was $5.78. September 12, 1967 was another cool and cloudy day. Again my father and mother travelled to Old Dubrovnik and took pictures. Wednesday, September 13th was sunny and warm, a fine day for relaxing at the beach Lapad. Thursday, September 14th, the morning hours were cloudy. They went gift shopping in downtown Dubrovnik and in the afternoon when the sun shone bright and warm, it was a perfect afternoon to relax on the beach. Friday, September 15, 1967, the day was dull, cloudy and it rained all day. They stayed in their motel and wrote letters to their daughter Joyce. September 16th the sun began to peer through the clouds and again my father with his Katharine went on to Old Dubrovnik. Sunday, September 17, 1967 was another cloudy and cool day. They went to Old Dubrovnik to see the aquarium. Monday, September 18, 1967 was a sunny day. They mailed a parcel to Canada, then took a bus to Budva; arriving in Budva at 4:00 p.m. They stayed at the Hotel Budva for the night and on September 19th they moved to the Hotel International then took a bus tour to Svati Stefan; a very interesting tour my father recorded in his journal; and they were at the beach in the forenoon. September 20th after an early breakfast they were at the beach and in the afternoon as the clouds rolled in they went to downtown Budva. Thursday, September 21st was a bright, warm, sunny day, they stayed at the beach all day and took pictures of the beach and the hotels, The Slavia and The International where they were staying.

Friday, September 22, 1967, their holiday in Yugoslavia had come to an end. My father with his Katharine at his side went into town to arrange their flight to Belgrade. They paid The Hotel International $60.00, and $20.00 for their airline tickets to Belgrade. Saturday, September 23rd, on a cloudy day they were leaving Budva by bus to Titograd and from there by airship to Belgrade. Their flight arrived in Belgrade at 11:30 p.m. and their accommodation for that night was the Hotel Palace. Sunday, September 24, 1967, after a quick and hurried early morning breakfast they were at the Belgrade Railway Station making reservations to travel by train to Lvov in the Ukraine and on Monday, September 25th they were at the Hungarian Embassy for their transit visas. My father recorded that he paid $9.00 for the visas and $67.00 for their railway tickets. Tuesday, September 26, 1967 they were on a train travelling on through Hungary to Lvov, Ukraine. They arrived in Lvov at 11:00 p.m. that night and took a taxi to Intourist to arrange their hotel; and on September 27th they went to pick up their train tickets to travel on. From the train station they sent telegrams to their family in Canada and to Marshall and Jo in Moscow. Thursday, September 28, 1967, they went on a tour and my father recorded in his journal "Katie's cousin came to meet us". In his journal he wrote "Katie was excited and happy to meet and to have conversation with a relative so far away from home".

September 29th was a cloudy day, cool and fine for travelling on and after a short visit with the cousin they were on their way to Ternopol at 4:00 p.m. arriving in Ternopol at 6:00 p.m. Saturday, September 30th they went on a tour of Ternopol. It was a bright, warm and sunny day, my father recorded in his journal, and he named two men they met, Bill Dzindra and Roman, Bill's son-in-law. Later that same day, both men arrived at the hotel suite to visit with my father and mother. (Dzindra was my father's mother's (my grandmother's) maiden name.) My father did not make a note in his journal if Bill Dzindra was related to him or not; or if the spelling of the name was similar to his mother's. – Some years later, I was told by my brother Patrick's wife Connie that they were not related. Connie said, and I quote "Your father travelled the world over and he did not find one ancestral relation". My father did say in his journal; they took pictures the day they met and in the evening they had supper together with their spouses in a restaurant with dance music; and they danced. Sunday, October 1, 1967 my father and mother alone, went on another tour of Ternopol, took pictures and went to a bazaar. Monday, October 2nd at 7:30 a.m. they were on their way back to Lvov, arriving in Lvov at 9:30 a.m. Father recorded in his journal that Zena, Katharine's cousin was there at the train station to meet them and that same evening they went with Zena to see a show. He did not record in his journal what the show was about, or where it was. October 3, 1967 they went site seeing, taking pictures, and to the Lenin Museum, with Zena as their guide. That same day they boarded an airship to Simferopol, and from Simferopol they went by taxi to Yalta – a health resort on the Black Sea; arriving in Yalta at 6:30 p.m. They took a suite in the Oreanda Hotel, and this hotel was their home to the 30th October 1967. Wednesday, October 4, 1967, my father and mother sent a telegram to Marshall in Moscow then went for a walk through a park and took pictures. October 5th was a bright and sunny day; a perfect day for an all day tour of all of Crimea. When they returned to their hotel suite, they found a telegram had arrived from their son Marshall, and a letter was there from their son Patrick in Canada. Friday, October 6, 1967 a letter arrived from Marshall in Moscow. Father immediately answered Marshall's letter and that afternoon, with his Katharine; they were relaxing on the beach and enjoying the Yalta sunshine. Saturday, October 7th with a soft breeze blowing through their hair, my father and mother were completely lost in the beauty and serenity of the sea, while strolling barefoot, hand-in-hand along the sandy sea shore enjoying the warmth of the sun shining down upon them. Sunday, October 8, 1967, they went walking along the beach taking pictures. When evening came they went to a symphony concert. A cloudy Monday morning October 9th did not stop my father and mother from going for a swim in the sea. In the afternoon when it rained, they stayed in their hotel suite reviewing the letters they had received from their sons; and they were resting and having a nap. Tuesday, October 10, 1967 was a sunny day. They went into town then for a stroll in the park. October 11th, they went into town to have Katharine's wrist watch repaired, father recorded in his journal; and in the afternoon they were sunning themselves; enjoying the heat of the sun. Thursday, October 12, 1967 they went for a stroll through the park and again were suntanning. October 13th they walked into town and went suntanning. Saturday, October 14, 1967 they walked into town to pick up Katharine's wrist watch from the repair shop. Sunday, October 15th was a warm and sunny day and all that day they were sunning and having a picnic in the park with a group of people they met in Yalta. They all were enjoying each other's company and having fun from dawn till dusk. October 16th they went into town and left that same wrist watch in a jewelry shop to be repaired. Tuesday, October 17, 1967, my father and mother walked into town and my father went for a complete physical examination with the doctors in the Yalta clinic. While there at the clinic they met people from Vancouver, British Columbia, Canada. Wednesday, October 18th my father had x-rays taken of his shoulder and of his ankle. He then had electric massage treatments done to both his shoulder and to his ankle at that clinic. October 19, 1967 was a bright, sunny day. My father went back into town with his Katharine, to have a second electric massage treatment and Friday, October 20th, though it was cloudy and it rained all day my father, with his Katharine at his side, went into town to the clinic for his third and final electric massage treatment. Saturday, October 21, 1967 was a warm, calm, sunny day. A great day to visit Chehov's Museum and the house where he used to live. They took pictures then went to Chehov's Theatre to see a concert. Sunday, October 22nd my father and mother went by taxi to the Nikitski Botanical Gardens. For most of the day they walked through the gardens, then took a boat cruise back to Yalta. October 23rd they went into town to relax in the park and to enjoy the warm sunshine. With just one week left before the 30th October arrives when they must say goodbye to the resort in Yalta; my father decided to go back to the clinic for one more electric massage treatment to his shoulder and to his ankle. When he and his Katharine entered the clinic, to their surprise, they found the clinic full of American tourists waiting inline for their electric massage treatments. While waiting his turn; my father and mother had a long talk and visit with the American tourists who made them feel closer to home. Tuesday, October 23, 1967, Marshall and Jo, arrived from Moscow for a three day visit. It was a beautiful sunny day in Yalta when they arrived and for their first day's visit they were walking and talking in the park the whole day enjoying each other's company, the clean air and the warm sunshine.

Wednesday, October 25th my father took his Katharine, his son Marshall and Jo on a boat cruise to the Nikitski Botanical Gardens where they took pictures and enjoyed the outdoor plants and flowers. Thursday, October 26th, they all went on a boat cruise for the day. Friday, October 27, 1967, after an early breakfast, my father took his family to Chehov's Museum, Chehov's house and the park before returning to the Yalta bus depot to see Marshall and Jo off on their return to Moscow. Saturday, October 28th was a cloudy day my father recorded in his journal. He went to arrange their boat cruise to Odessa and gave notice to the Hotel Oreanda for Monday, October 30th while his Katharine was at the hair dressers getting a permanent curl. Sunday, October 29th, was another cloudy day, my father recorded in his journal that they stayed in their hotel suite packing their suitcases and writing letters and post card to people they met during their travels and to their family back home in Canada. Monday, October 30, 1967, the day had come when they must leave. A taxi delivered my father and mother to the ship Ukraina and they sailed away on an overnight cruise to Odessa. October 31, 1967 their ship Ukraina docked in Odessa at 8:00 a.m. First they went to update their passports then on an all day tour, and to see a circus. November 1, 1967 Wednesday, they went site seeing and taking pictures the whole day. The morning hours were cloudy Thursday, November 2nd, my father and mother went for a walk through the Odessa park and took pictures, then they were off to the airport where they had lunch, and at 3:45 p.m. they boarded an aircraft to Kiev, arriving in Kiev at 5:10 p.m. From the airport they took a taxi to the Intourist Hotel. Friday, November 3rd , was a cloudy day in Kiev. From their hotel, my father and mother walked to Krischatic Street to a department store, then to a theatre to see a movie. Saturday, November 4th my father with his Katharine went on an excursion of Kiev. They saw Picharska Larwa, Krischatic, and the Dnieper River, where they stood on the river's bank and watched the mighty river flow down to the Black Sea.

Sunday, November 5, 1967 was a cloudy day in Kiev, they took pictures of Sophickey Sobor Church, and on Monday, November 6th, on another cloudy day, they went by hydroplane along the mighty Dnieper River to Konev Hill where the monument of poet Taras Shevchenko, a Ukrainian Bard, stood, and there on Konev Hill; they went through the Taras Shevchenko Museum, took pictures, then returned to Kiev by bus. Tuesday, November 7, 1967 my father and mother went to Krischatic Street to see a parade and they took pictures. Father recorded in his journal "there were thousands upon thousands of people there celebrating". He did not say what they were celebrating. Wednesday, November 8, 1967 they went by subway to "Georges Place" for supper and on Thursday, November 9th at 5:30 a.m. they took a taxi to the Kiev airport. It was foggy at the airport and their flight did not leave the airport till 2:30 p.m. arriving in Sochi at 5:30 p.m. From the airport they went by taxi to the Intourist Hotel – a hot springs hotel. Friday, November 10th and Saturday, November 11th, though both days were sunny days, my father and mother stayed in the hotel soaking in the hot springs hot tub. Sunday, November 12th, they went for a quick walk in the warm sunshine and soon were back in the hot springs tub. Monday, November 13 they went for a long walk and on Tuesday, November 14, 1967 they went on a bus tour all around Sochi. Wednesday, November 15th they walked into town where father bought a cap and they went to see a movie. Thursday, November 16, 1967 they went to a botanical garden park. Father recorded in his journal "the park was beautiful and it was a beautiful sunny day". Friday, November 17th another sunny day, perfect for sunning and walking along through the beautiful botanical Sochi Park. Saturday, November 18th though it was raining and it was a dull and cloudy day; they went walking in the rain; getting soaking wet; knowing that the mineral hot tub bath was there waiting for them when they return. Sunday, November 19th more rain, another dull cloudy day and another day to enjoy the hot tub. Monday, November 20, 1967 was a bright and sunny day. First; my father took his Katharine for a long walk outdoors and when they returned to the Hot Springs Hotel they packed their suitcases for tomorrow they must be moving on. Tuesday, November 21st at 9:00 a.m. they boarded an aircraft to Tbilisi arriving at the Tbilisi airport at 11:00 a.m. From the airport they went by taxi to the Intourist Hotel. Wednesday, November 22, 1967 they went on an excursion tour of Tbilisi, taking pictures. Most interesting my father recorded in his journal; was being inside a Sioni Cathedral of the 5th century while a church ceremony was in progress. Thursday, November 23rd my father and mother received a telegram from Vernon, British Columbia, and immediately replied to the telegram then they went walking through the town of Tbilisi to see the shops and to talk with the people. November 24, 1967 my father with his Katharine were off to the airport early that morning and boarded an aircraft to Ashkhabad, arriving in Ashkhabad at 2:40 p.m. Saturday, November 25th they went on an excursion to see a rug and carpet manufacturing company and to a museum, then they boarded a flight to Tashkent, arriving in Tashkent at 5:00 p.m. Sunday, November 26, 1967, they went walking through all of Tashkent then boarded a flight at 4:15 p.m. to return to Kiev, arriving in Kiev at 9:30 p.m. From the airport they took a taxi to the Hotel Dnipro. Monday, November 27th they were up at daybreak and looking out their window they found it had started to snow. They stayed in their hotel room and wrote a letter to Patrick and Connie in Vernon, and many postcards to friends they met along the way. Tuesday, November 28th, they went walking and window shopping along Krischatic Street then again they went to see the Taras Shevchenko Museum. Wednesday, November 29, 1967, the snow was snowing all over Kiev. My father with his Katharine went shopping for overshoes then took the subway to "Georges Place" where they stayed for the whole afternoon and watched the huge white snowflakes softly falling down. Thursday, November 30th a telegram arrived from Patrick in Vernon to which father immediately sent a reply then sat down and wrote Patrick a long letter, after that they went to "Georges Place" for a last visit. December 1, 1967, a Friday and still snowing in Kiev my father with his Katharine stayed in their hotel room for the whole day packing their suitcases. Saturday, December 2nd, my father recorded in his journal "Roman came to our hotel room to see us before we left for the airport, where we waited for 3 hours for our flight to Moscow". He did not say who Roman was. They arrived in Moscow at 5:30 p.m. where they got a room, No. 411 in the Intourist National Hotel. Sunday, December 3, 1967, snowing in Moscow, my father made a telephone call to Marshall and Jo and that evening they all went to see a hockey game: Canada playing the U.S.S.R. Father did not say who won the game. Monday, December 4th still snowing in Moscow. With their Kiev overshoes on, they went walking around the Red Square then they went to see the St. Basilica Church. Tuesday, December 5th, more snow coming down all over Moscow, my father with his Katharine went inside the Kremlin and into the museum. Wednesday, December 6, 1967, still snowing with no signs of stopping, my father and mother went walking into town and bought top coats and overcoats then went to the "Palace of Congress Theatre" to see the Ballet Cinderella. Thursday, December 7th my father and mother went site seeing around Moscow, then they went shopping. Father wrote in his journal "Mother went shopping" after shopping they went to the Moscow University and there they enjoyed the evening hours at the University dormitory visiting with Marshall and Jo. Friday, December 8, 1967 after shopping in a General Store in the Kremlin Square, my father and mother went to the Satira Theatre to see a drama. Saturday, December 9, 1967, with Marshall as their guide, they all went site seeing for the whole day and when evening came, my father with his Katharine, Marshall and Jo, were out dining, dancing and exchanging conversation, late into the night.

December 9, 1967 was the last day my father recorded in his travel journal. Not another word was said. He did not say when they left Moscow or how. December 15th was my mother's 59th birthday, and my father would have made sure her birthday was celebrated, though he did not say where. He purchased a gold watch for his Katharine; a souvenir U.S.S.R. gold watch, this watch was for her birthday.

With their Kiev overshoes on, and their Moscow topcoats and overcoats; my father and mother returned to their Apple Valley home in Vernon, British Columbia, just before Christmas. December 1967. All the family got together in their big family home on Mission Hill in Vernon to celebrate Christmas, and to enjoy a turkey dinner with all the trimmings. Down stairs in the family room were presents for everyone under a beautifully decorated Christmas tree.

Chapter 32

There's No Place Like Home

After my father's travelling years were over; he was approached by a Vernon newspaper reporter for a story of his world travels; His words were: "There's no place like home". Those few words summed it all up. The home he created and his family were his domain that held the most value and nothing, no-one and no-where could ever replace.

Mid pleasures and palaces though we may roam
Be it ever so humble, there's no place like home:
A charm from the sky seem to hallow us there,
Which seek through the world, is ne'er met with else-where,
Home! Home! Sweet Home. There's no place like home.
There's no place like home.

Reading between the lines of my father's 1967 travel journal through the U.S.S.R. Europe and Asia; I sensed a sadness; and that sadness, I feel was that he did not find his ancestral roots while travelling and searching the world over. He must have felt very alone and very far away from his Canadian home. Had he travelled on with his Katharine; from Tashkent, on into Upper Mongolia then explore the Yablonovy

Mountains and the Yenisey River Valley, perhaps there he may have found some knowledge of his ancestry: Though it was the wrong time of year to travel on. Winter was approaching and soon every country not explored by my father would be covered white with snow.

During the 1800's, as during the 1900's people from far and near settled in countries where they could find work or get set-up in a business to make a life worth living for themselves and for their families. My father's grandson Lorne: Edward's first born son, wandered away from his Canadian home to Bangkok, Thailand, and there he stayed for 3 to 5 years working for the Government. From Bangkok he was transferred to the Philippines where he met and married a Philippean princess. The year 1997, Lorne returned to British Columbia, Canada with his bride where they are building their home and their life together. What made my father's grandson Lorne go to Thailand, then to the Philippines? What was calling him and beckoning him to leave his country Canada to a land unknown? That is the mystery of life ancestral generations follow.

1970, my father with his Katharine travelled on to see South America and Mexico. Their baby daughter Joyce, who was now 28 years old arranged this excursion to travel and went along with them to visit Venezuela, Brazil and many other countries in the South American continent. Travelling to South America, all tourists had to receive an immunization vaccine for yellow fever. My father must have suffered a viral infection when they entered South America and their tour of South America was cut short. After visiting Brasilia, the capital city of Brazil, they flew to Guatemala then to Mexico City in Mexico. In Mexico City my father, his Katharine and their daughter Joyce hoped to recover strength there to continue their South American tour when my mother got ill. The Mexican food did not agree with her and their agenda to visit other countries in the South American continent was cancelled. From Mexico City they flew to San Francisco; to Seattle; then back to their Sunny Okanagan Valley home. To continue travelling South America; their interest slowly diminished and they did not persevere.

Spring 1972, my mother was away in New York, visiting her sister Agnes when I arrived in Vernon. There, in this beautiful Apple Valley I found my father home alone. I packed a suitcase for him and asked him to come to Vancouver and together we would fly to Arizona to watch his grandson William and the Ogden Utah Wildcats football club play Flagstaff Arizona. I then arranged a flight, and sent a ticket to New York to my mother, to fly across the U.S.A. – East to West and to meet my father and me in San Francisco. When my father and I arrived in San Francisco and found our way to the airport terminal; there we found my mother waiting for us: She had arrived a half hour before we did; and there she was sitting in the San Francisco Airport waiting room, wearing her fur coat. We rented a car in San Francisco then drove the California coastline to Los Angeles. From Los Angeles we flew to Phoenix, Arizona where we rented a motel-hotel suite in a big new building in Scotsdale surrounded in beautiful park-like grounds, with palm trees and tropical flowers. An excellent restaurant and a swimming pool with a Jacuzzi pool at one end of the swimming pool. The day the Wildcats were arriving in Flagstaff; we drove the Arizona Mountains to Flagstaff. A very enjoyable, beautiful, cool mountain, fresh air and green drive, it was away from the hot and dry Arizona desert. After we watched my son William and his football team – the Ogden Utah Wildcats play Flagstaff Arizona, we had a short visit with him and after the football team flew back to Ogden, Utah; my father, mother and I travelled on to Tucson, arriving in Tucson right in the middle of a Kirk Douglas movie being filmed. Immediately we were sent into a restaurant to be a part of the restaurant crowd while the shooting; and on horseback Kirk Douglas galloped by. There was a lot of excitement there in Tucson that day; on the streets, in the restaurant, in the gift and souvenir shops, the bank, the post office and in all the buildings that were set-up for that movie. While in Tucson we went on tour of a Dean Martin Spanish movie set-up. All the barns with all the horses were there and the surrounding area; the shops, the huts and the street, all looked very Spanish. Though Dean Martin was not there, or; he may have been hiding somewhere with his bottle of wine. In the Arizona desert are native Indian villages: A native bakery, and gift shops with beautifully handmade turquoise jewelry on display. Arizona is also; fields and fields of orange and grapefruit groves and acres of cotton. We stayed in Scotsdale for a few days after our tour of Arizona was complete to swim in the swimming pool and to relax in the whirlpool, before we travelled on to see the Grand Canyon. There to see the grand Canyon were many tourists and many of whom rode the mule train along the very steep banks, down to the Colorado River below. We were told that mules and mules only had the footing and were able to scan that very steep walk from above down to the Colorado River below. We could see some gardens and a few homes along the Colorado River though a very little bit of sunshine could reach them. My father, mother and I took an aeroplane, and flew over the Grand Canyon. We were able to see it all as the pilot banked the aeroplane from side to side. After the Grand Canyon we continued on to Salt Lake City, Utah. The Salt Lake was interesting to see from the aircraft, the colour of the Salt Lake from above was a deep red and purple colour and much of that Salt Lake had already dried up leaving shores of salt and sand. The gift shop at the lake had many little creatures on display that even now; still live in the Salt Lake. My son William with his 1972 GMC Jimmy was there to meet us when we arrived at the airport. From the airport, he took us on tour of the Salt Lake then through all of Salt Lake City. The most interesting to see in the city; was the Mormon Church with a golden seagull perched on top of a pedestal which stood as high as the church steeple. The seagull has been a most valued bird in Utah since the pioneer years. The story is; that when the pioneers first arrived in Utah in 1847, the grasshoppers arrived there the following spring. They were so huge and very many of them, they overpowered every living thing. Then came flocks of seagulls and the grasshoppers disappeared; that is when life first began for the pioneers and the seagull became a Mormon legend. A monument was erected on Temple Square in commemoration of this great life-saving event. We were allowed to visit the Mormon Church Tabernacle with its huge pipe organ. That pipe organ was enormous, and that was the Mormon Tabernacle from where the famous Tabernacle Choir sang so beautifully and recorded their songs for the whole world to hear; And they still do. The church temple was kept private for Mormon members only. On leaving Salt Lake City we travelled on to Ogden; then we entered a small canyon and drove along a gently flowing river with beautiful homes on the other side of this river with their rock gardens built right down into the river bed. This was a beautiful cool drive. At the far end of this canyon where the mountains widened was a famous excellent restaurant "Maddocks", and we arrived there right on time to enjoy a delicious dinner. My father, mother and I stayed in Odgen for a week in a Holiday Inn Motel, near the Weber State College where my son was continuing his education on a football scholarship. The motel was near the football stadium where we walked to, and watched the football practices every evening through the week. On the week's end we travelled by bus to Provo, to watch yet another football game: The Ogden Utah Wildcats playing Provo. People in the U.S.A. are true football fans. The stadiums were always full and full of excitement; with enthusiastic cheering and shouting from the beginning to the end of the games. The following morning after the football game in Provo, was a Sunday and my father, mother and I were preparing to return to our Canadian homes. My mother and I were packing our suitcases while father went to the Holiday Inn Motel office to pay for our meals and our suite for the week. William, my son, was on his way from the college for a short visit then to transport us to the airport. With a surprised look on his face, my father soon returned from the motel office and said, "There was NO CHARGE", our invoice was paid in full. I accompanied my father to the Football Club's Office and when my father approached the desk of Coach Mike Caviness; Coach Caviness said to my father "The football club agreed to pay for your stay at the motel, and we hope you enjoyed your visit in Ogden; and enjoyed being at our football games. This is a very small offering compared to the value our football club receives from having your grandson play on our team. Your grandson plays any position in offence and in defense and he plays them with enthusiasm and strength. He often plays both sides during a football game, filling in for a player who gets hurt, or is unable to play. We have not had a player on our team who was able or willing to play both sides during my many years in football career"; Coach Mike Caviness said.

And he told my father and me that Bill is the first young man he has ever recruited to their college football team, who is there to study as well as to play football. He said "Most young men sign up for the game only: Bill has a full college schedule and he is a top of the class student." My father thanked Coach Mike Caviness for his sincerity; then shook his hand and we left. When we arrived at the motel suite; William was there visiting with his grandmother and ready to deliver us to the airport. It was in the springtime 1976, when William returned to Canada: A professional athlete, with a college degree: A bachelor in Science, Psychology, and Business Administration. I put together several scrap books and albums of most of the Odgen Utah Wildcat football games while Bill was a Weber State College student. Of where the games were played, of the programmes, and of which team won the game. After William's college years were over; I gave him those albums and scrap books to remind him of the days and the memories;

The "Student Prince" William Alexander Frederick during his college years.

When he was a "Student Prince".

A game is a trial of skill or strength for sport or amusement: And Life

It is not if you win or lose…

But how you play the game.

Being away from home in a college; meeting, sharing, and experiencing a new and exciting college way of life is knowledge. A knowledge that is not learned by being at home only: Travelling the world over; seeing and being in many countries, having many friends and acquaintances is a knowledge that is not learned by being at home only. This knowledge that is absorbed by being away from a native home and family is a vainglory experience to be treasured. The most precious, with the most value that nothing, no-one, no-where could ever replace is a father, a mother, a son, a daughter: A family and home – however humble. My father stressed upon these values: They were of his dearest and best.

An exile from home, splendor dazzles in vain.
Oh, give me my lowly thatched cottage again;
The birds singing gaily, that come at my call;
Give me them, with that peace of mind, dearer than all.

To thee, I'll return, overburdened with care,
The heart's dearest solace will smile on me there.
No more from that cottage again will I roam.
Be it ever so humble, there's no place like home.
Home! Home! Sweet Home – There's no place like home
There's no place like home.

A repayment in appreciation to Coach Mike Caviness for his generosity towards my father, mother and me during our visit to Ogden Utah in 1972: On July 22, 1976 after the B.C. Lions football camp in Penticton was over, and my son William's (Bill Norton's) first Canadian professional football game was set to be played at the Empire Stadium in Vancouver: I sent a flight ticket with all expenses paid to Coach Mike Caviness in Ogden Utah, to be in Vancouver, and to be at Bill Norton's first Canadian professional football game. With only three downs, from a spectator's point of view, Canadian football is a lot more interesting to watch than American football with four downs. During the years I watched football – American and Canadian – many times there were more than three or four down; there were a pile of them down many times. This was a first time visit in Canada for Coach Caviness and a first time for him to be at a Canadian professional football game. On leaving Canada, Coach Caviness said that his visit to Vancouver was an entertaining adventure and a pleasure. During his college years 1972 – 1976, in Ogden Utah; my son William together with Coach Mike Caviness were friends with much respect for each other and I was pleased. I was also pleased that coach Caviness was able to attend William's first Canadian professional game. Football is a powerful contact game and I was always concerned. Concerned through all of my son Bill Norton's football games, in college, and through all of his professional football years.

B.C. Lions - 1976

Head Coach - Cal Murphy
General Manager - Bob Ackles
Sports Director - Denny Boyd

B.C. LIONS

4	*MIKE STRICKLAND	RB
6	*JOHN SCIARRA	QB
8	ERIC GUTHRIE	QB
10	BARRY ARDERN	DB
11	*VINCE PHASON	DB
12	*BOB PEARCE	DB
14	*JOE FOURQUREAN	DB
16	*ROCKY LONG	DB
18	LUI PASSAGLIA	PK
20	ROSS CLARKSON	FL
21	BARRY HOULIHAN	RB
25	LEN PLATT	SE
27	TED DUSHINSKI	DB
28	DOUG CARLSON	DB
30	JIM YOUNG	SE
31	*LOU HARRIS	RB
32	*GRADY CAVNESS	DB
35	*JIM BAKER	LB
36	TERRY BAILEY	RB
37	GLEN JACKSON	LB
50	BRIAN SOPATYK	G
51	*RAY NETTLES	LB
52	AL WILSON	C
55	*EDGAR BELL	G
58	*NEAL SKARIN	DT
60	*DON WUNDERLY	DT
62	*LAYNE McDOWELL	T
64	*ALLEN GALLAHER	T
65	*JIM SCHNIETZ	G
→ 66	BILL NORTON	G
67	GARY ROBINSON	DT
72	*MERCED SOLIS	TE
74	SLADE WILLIS	SE
76	BILL BAKER	DE
77	WAYNE SMITH	DE
78	PAUL GIRODAY	LB
79	*DAN McDONOUGH	LB

*Denotes import

SASKATCHEWAN ROUGHRIDERS

10	*TED PROVOST	DB
11	DAVE SYME	QB
12	*RAY ODUMS	DB
14	RANDY GRAHAM	DB
15	BOB MACORITTI	PK
17	LORNE RICHARDSON	DB
19	STEVE MOLNAR	RB
20	KEN McEACHERN	DB
21	AL FORD	TE
23	RON LANCASTER	QB
24	*TOM CAMPANA	RB
25	*RHETT DAWSON	FL
26	*JIM MARSHALL	DB
27	*PAUL WILLIAMS	DB
31	LOU CLARE	LB
33	*PETE VAN VALKENBURG	RB
37	*ROGER GOREE	LB
42	LARRY BIRD	C
52	JIM HOPSON	G
53	GARY BRANDT	T
58	*SAM HOLDEN	G
59	*RALPH GALLOWAY	G
60	RON CHERKAS	DT
61	*JESSE O'NEAL	DE
62	FRANK LANDY	LB
64	*MIKE DIRKS	DT
65	*BRIAN VERTEFEUILLE	T
66	*TIM ROTH	DT
69	*GEORGE WELLS	DE
70	STEVE MAZURAK	SE
71	LEIF PETTERSEN	SE
72	*CLEVELAND VANN	LB
75	BILL MANCHUK	LB
71	BOB RICHARDSON	TE

*Denotes import

Thursday, July 22, 1976

№ 9680

Professional football is not a game – a trial of skill or strength for sport or amusement: professional football is MEAN, MONEY and POLITICS. Bill Norton quit the B.C. Lions: he walked off the field never to return to the B.C. Lions football club because of their cruel policy towards Canadian players. From the day Bill Norton returned from college; all day and every day our telephone did not stop ringing. Every Canadian football club across Canada were after Bill Norton to join their team. They were offering Bill a player position of his choice with excellent pay. The B.C. Lions would not release Bill Norton to any other team. Head Coach, Cal Murphy and General Manager, Bob Ackles said, "If Bill Norton does not play for the B.C. Lions, we will make sure he is benched for the year and does not play for any other team in Canada: — that they would disable him – and they tried. With a one-on-one. Cal Murphy and Bob Ackles were after Bill to play his first year for B.C. with no pay, Coach Caviness set them straight on that before he returned to Ogden, Utah. Bill Norton played for the Calgary Stampeders for a few months in 1976 while waiting to be transferred to Toronto, to play for the Toronto Argonauts. 1982, Bill went to play for the Montreal Alouettes when they needed a lift. My first granddaughter Lacey Genevieve, was born in Montreal, September 5, 1982, while my son was playing for the Montreal Alouettes. Before Bill Norton left Montreal, the Alouettes had re-named their team "The Montreal Concords". After the Concords were on track, Bill returned to be with the Toronto Argonauts, (his home team) and he helped them into victory of winning the Grey Cup in 1983. They played the B.C. Lions.

My second granddaughter Larissa Nicolette was born February 10, 1988 after my son William retired from football.

Grey Cup 1983

B.C. LIONS

5	LUI PASSAGLIA	K
6	*JOE PAOPAO	QB
10	JOHN BEATON	LB
11	*LARRY KEY	RB
12	*MIKE NOTT	QB
14	*JOE FOURQUREAN	DB
16	*DEVON FORD	DB
17	*FRANK OLIVER	DB
18	*KEN HINTON	DB
19	DON TAYLOR	RB
20	*JOHN HENRY WHITE	RB
25	AL CHARUK	WR
26	*LEON BRIGHT	WR
28	DOUG CARLSON	DB
30	JIM YOUNG	WR
36	TERRY BAILEY	WB
37	GLEN JACKSON	LB
40	PAUL GIRODAY	LB
44	RON MOREHOUSE	LB
45	*SAM BRITTS	LB
52	AL WILSON	C
54	GLENN LEONHARD	OG
55	*JIM LOHMANN	OT
58	NICK HEBELER	DT
60	*DREW TAYLOR	DE
61	DOUG SEYMOUR	DT
64	*LARRY WATKINS	OT
68	JOHN BLAIN	OG
69	JOHN BLAKE	OG
70	MARK HOUGHTON	WB
72	*HARRY HOLT	TE
77	*LOUIE RICHARDSON	DE
78	FRANK LANDY	DT

*Denotes Import

TORONTO ARGONAUTS

7	*M. L. HARRIS	TE
8	*MARK JACKSON	QB
9	IAN SUNTER	K
11	*TONY ADAMS	QB
15	*JIM MARSHALL	DB
16	*ERIC HARRIS	DB
19	*MIKE KRAMER	DB
20	PAUL BENNETT	DB
21	*TERRY METCALF	RB
23	VIC ANONSEN	WR
24	MARK BRAGAGNOLO	RB
25	PAUL PEARSON	WR
27	*MIKE McARTHUR	RB
28	PRESTON YOUNG	DB
30	*BILLY HARDEE	DB
33	DWIGHT EDWARDS	WR
34	BILL PALMER	LB
35	*RICHARD LEWIS	LB
39	DAN HUCLACK	TE
50	KEVIN POWELL	OT
51	*MIKE OBROVAC	OG
53	*RICK RAZZANO	LB
54	AL MacLEAN	OT
56	JOHN FOUBERT	DT
58	*BRUCE KIMBALL	OG
61	*BRUCE SMITH	DT/DE
62	*ECOMET BURKLEY	DT
68	DOUG REDL	C
69	BILL NORTON	OG/OT ←
71	RON SOUTHWICK	LB
72	PETER MULLER	TE
77	WAYNE SMITH	DE
79	JIM CORRIGALL	DE

*Denotes Import

#69 - Bill Norton Offensive guard and tackle

Lacey

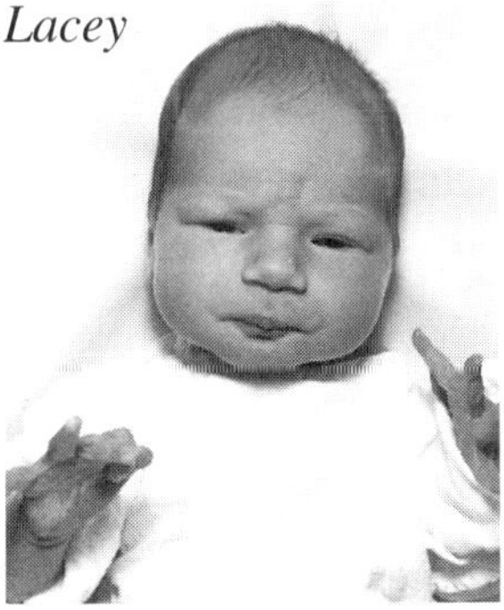

Larissa

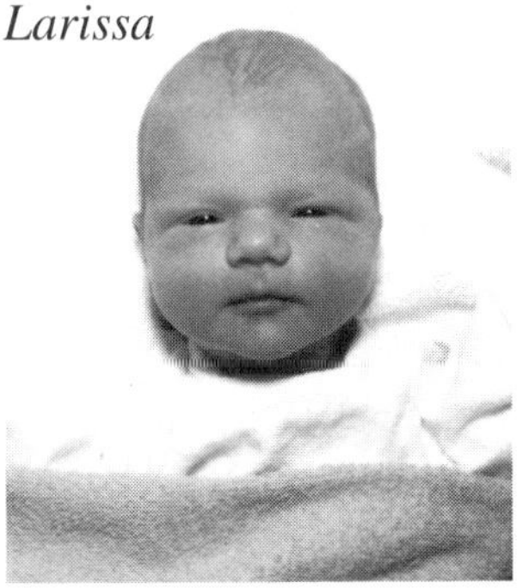

Chapter 33

Jet Trails in the Sky

There is a destiny which makes us brothers
None goes on his way alone
All that we send into the lives of others
Comes back into our own.

These are words by Edwin Markham: Does he speak of KARMA; I believe so. Of the animal world: People are the most intellectual, cruel, selfish, destructive; and there are the kind, caring and giving – A fascinating lot; are people. **In BRIEF MEMOIRS OF MY FATHER:** I said; "City taxes on his building "The B-Y Store" went up so high he could not handle and sold out." — These are words I heard about my fathers B-Y Store from people in and around Vernon. Envious people who arrive at a conclusion; a conclusion of their own making then believe it to be true: Be it family, be it family physician, be it client, a customer and so called friends. The attitude of these people is: — My mind is made up – Do not confuse me with facts. And you cannot; you cannot change their thoughts with facts. My father knew that and he did not even try. When words were said that the B-Y Store was sold because the city taxes were too high for my father to handle; he said nothing; he knew there would be no point in saying anything. He kept his failing strength, and the reason why he sold his B-Y Store in complete confidence. He did not want pity from anyone, nor anyone to feel that no longer they could rely on him – on his strength. Serious matters are better left unsaid; no-one wants to listen – and to hear.

The 6.3 acres on Scott Road in Vernon. The weeping willow tree beside the house.

House in the background. Mother and Father in corral.

Corral and grassy meadow.

Father with his cows on 6.3 acres

Father and grandson Bill with a cow on 6.3 acres

Mother and father in front of their house at 3407 - 18th Avenue, Vernon

My father's main concern was to make sure his sons have an easy start in their adult private lives; and he produced: For his youngest son Marshall James, he purchased the 6.3 acres, on Scott Road in Vernon, for when Marshall and Jo return from the U.S.S.R. and the Moscow University, they would have a home to come to. Patrick with his family had the 10 acres on Okanagan Landing Road. For his son Henry, Joan and their 5 sons and 3 daughters who were renting a house in Grimsby, Ontario; my father purchased 50 acres of land with a house and barns in the Caistor Township of Niagara Falls. Now Edward; a first born son; Why! Edward got it all: He was like a prince of kings. My father's happiness was looking after his family's needs. He never thought of going off alone to play a game of golf. Or to sit on a river's bank and listen to the quiet of the outdoors, the song of a bird and the rippling sounds of the river while fishing, or trying to catch a fish. To relax; to get away from family demands, he did not do. Whatever he did or planned was always for his family; with his family.

During the summer of 1974, my father took his Katharine on a month's long bus tour of the British Columbia coast – North to Alaska, then to the Yukon. Leisurely and relaxed they travelled along in a bus full of seniors; and they were stopping to see all the historical sights, along with interesting and exciting stops for their lunches and for their overnights. When they returned to their sunny Okanagan Valley home; they were, relaxed and happy. Joyce; their youngest; their baby, married a handsome Curt Hoffman in August 1976. She was the last family member to leave their home. With her away, the family home seemed very empty leaving my father and mother all alone, though not for long. In the autumn of 1976 they flew to visit Cuba for two weeks and because they enjoyed their visit in Cuba; the hospitality, the service and the food were excellent they said; they went again for another two weeks in 1981 after their 55th wedding anniversary celebration. This time they went to accompany my mother's youngest sister Jenny on her first visit to Cuba.

By the year 1984; not only did my father have to bear his failing strength and the pain he suffered; his eyesight had diminished to the point he was not able to take his Katharine for a drive in the country or to an evening movie. Doctor Richard Story, B.Sc., M.D., F.R.C.S. (c) Ophthalmology, who's office was in Vernon said in his report "Night-time driving is not to be under-taken under any circumstances". And by 1985 my father had to give up his driver's licence and driving his car altogether. This must have been as difficult for him to accept as the pain he was suffering. A man who everyone relied on, a man who was strong and stood alone and on his own through all of his life, now had to become even stronger yet – to have to start relying on someone else and to be able to accept such a serious blow to his developing weary life. Doctor Richard Story referred my father to Doctor Gordon S. Harris, Retinal and Vitreous Referrals, in Vancouver; who then referred him to Doctor Eric L. Smith, Ophthalmologist also in Vancouver. It was the year 1988 when Doctor Richard Story of Vernon requested yet another opinion and sent my father to visit Doctor Gary Servais, Ophthalmologist in Vancouver. This letter from Gordon S. Harris explains my father's vision in detail.

GORDON S. HARRIS
B.A., M.D., F.R.C.S.(C)
RETINAL AND VITREOUS REFERRALS

EYE CARE CENTRE
2550 WILLOW STREET
VANCOUVER, B.C. V5Z 3N9
TELEPHONE (604) 875-4243

1984 08 10 10 Apr 84

Dr. Richard Story
P.O. Box 1208
Vernon, B.C.
V1T 6N6

Dear Dick:

Re: Mr. William B. YAWNEY

Thank you very much for asking me to see this 79 year old gentleman. He has had poor vision in his left eye for the past several years but since Christmas he has noted some distortion and decrease in visual acuity in his right eye. There is no history of flashing lights. He can recall no injury and there is no family history of eye disease. He enjoys good general health. He denies allergies and is on no medications. He has had floaters in his left eye for the past year.

His visual acuity with his slight correction is 20/40-2 in each eye. No improvement occurs with pin hole. Applanation tensions are 20 mm Hg. bilaterally. His anterior segments are unremarkable.

He shows a large detachment of the retina and pigment epithelium at and immediately temporal to the right fovea. There is no evidence of hemorrhage or subretinal cholesterol associated with it. The optic nerve head is unremarkable. The left fundus shows a similar lesion at and immediately temporal to the fovea, and measures some 3 x 5 mm in size. Some subretinal hemorrhage is present. No definite patches of neovascularization can be seen in either eye. However, I have performed Argon photocoagulation to both of these disciform lesions (detachment of the retinal pigment epithelium) in the hopes that this may cause it to flatten and improve his visual acuity. I have given him a 25-50% chance of success.

Thank you very much for asking me to see this pleasant gentleman.

Kind regards.

Yours sincerely,

Gordon S. Harris, M.D.

GSH:LW

cc: Dr. H.L. Clarke
Sim Wee, R.N.

Why? And what is it that causes so much pain for some to have to bear. Could it be that the very cold; the 60° below zero weather in Alberta my father worked outdoors in without proper clothing to keep him warm; and while he drove his team of horses 14 miles into our Viking town in the wintertime. Is that what did so much damage to his eyes. Could it be. If doctors would only listen to their patients, then they may find what the cause of a problem is. Today! They do not listen, or if they do; they do not hear; they give you a pill and send you away. If you can find a doctor who truly listens and cares; and respects his profession, never let him go. They are the few and far between. I know of only one such doctor since the 1950's, his name was Doctor Anthony Yurkovitch, one of the greatest and best. It seems to me that Doctor Gordon S. Harris is quite thorough in telling us what he found that distorted my father's eyesight; not why his sight was distorted – or what could be the cause.

1986, after my father and mother celebrate their 60th wedding anniversary held in the Vernon Lodge Hotel with all the family and guests from all over Canada and the U.S.A; my father took his Katharine on an anniversary ocean cruise to Alaska. They did not have to get off the ship but cruise along and enjoy the month of June sunshine with the fresh ocean breeze to blow their cares away. In the autumn of 1988 they went a second time cruising to Alaska; this time they went in the late autumn to see the British Columbia coast with all her beautiful autumn colours and to see the snow – and dog teams in Alaska! Returning to an empty family home.

Tired and alone with his beloved Katharine and memories of the past; he said to her "There were just the two of us when we started". – Remembering their golden days:

Golden days in the sunshine
of our happy youth
Golden days full of innocence
and full of truth
In our hearts we remember them
all else above
Golden days, days of youth and love.

I returned one day and found a lonely man, feeble and tired: This man I knew of power and strength with a heart of gold, was all alone and afraid in a Vernon Polson Acute Care facility, like a lone star in a galaxy. He seemed so all alone though this care shack was crowded with many; a variety of a many abandoned broken hearts; troubled minds and crippling, aching, tired bodies. It was lunch time when I arrived and found my father sitting at his table trying to have his lunch not seeing and not knowing what it was. I stood there for the longest time watching my silver haired daddy with tears streaming down my face, choked, I could not speak: I could not believe, I was numb.

In this vine covered shack in the mountains
Bravely fighting the battle of time
Was this dear one who's life withered sorrow
This silver haired daddy of mine.

My father was so afraid of a hospital; never through all of his life had he ever stayed in a hospital or had to have the care and attention of doctors and nurses. Why! He was born in the Canadian Northern wilderness in a cabin with no-one else but his mother and the howling sounds of timber wolves outdoors. Now! Because he could not see he was begging his Katharine and his sons to take him home. They did not. To visit his home for a few hours midday was all that they allowed. My father's Katharine and their son Edward said that father would be too difficult for them to handle if he was to remain in his home. With the money he still had, and the money he left in his will; he earned the best male nursing care for every day and every night in his own home for all of the remaining days of his life –which he did not get. My father always said; believed; and lived, through all of his life "To do unto others as you would have them do unto you." Can you imagine being abandoned during ill health and what this was doing to my father? I could see his abandoned feeling reflecting in the sadness of his dim brown eyes: And, there was nothing, nothing at all that he could do; and nothing that I could do either. His Katharine and their son Edward whom he trusted and relied on; he could not convince to take him home. They were the two who held the power. His power of attorney. Every day while in this old vine covered shack when his Katharine was not there beside him; he always asked "How is your mother; is she alright; I do want to see her. Would you ask her to come."

April 30, 1992 was an Easter weekend and my father's 87th birthday. His youngest son Marshall James brought him to his home from the Polson Acute Care Centre to celebrate his birthday with his family; his sons, daughters, grandsons and granddaughters were all there. My father was very feeble though he showed appreciation and was happy to be in his own home with all his family there beside him. This 30th day of April 1992 was a beautiful warm sunny day with a clear blue sky above. On this huge sundeck of my father's home, with the view of his sunny Okanagan Apple Valley all around us; we set up a buffet with a birthday cake and candles for my father and had chickens spinning on a spit in the barbecue. In the air a slight breeze was blowing and birds of every variety were chirping and singing in the trees my father had planted in his yard and garden. A family of swallows were nesting in the carport beneath the sundeck. They were swooping in and out of the carport catching flying bugs in the air. Then they would sit on the clothesline and watch us. A curious, interesting little bird is the swallow. My father could hear the birds singing though he could not see them. He could not see too clearly his family who were all there with him, though from the sounds of their voices, he knew they all were there. An amazing surprise to us was that father could clearly see jet trails in the sky, and he knew where they were going to or coming from. Jet trails so very high up in the sky.

April 30th, 1992 Father's 87th Birthday celebration with his Katharine, their four sons and three daughters.

May 1 1992, Father's visit in his home from 1:00 to 3:00 pm.

The next day; the 1st day of May 1992, again my father was brought to his home in the afternoon to visit with the family who stayed over an extra day and with his neighbours, friends and old acquaintances who gathered at his home that day to be with him for conversation, cake and tea. I was ready to leave that afternoon, to return to Vancouver Island when my father asked me to stay another day. He said he had some money he had saved over the years for his boxcar girl and he wanted me to have this money before I leave; that tomorrow he would ask Mom and Ed to get the money for me: That tomorrow never came; after Edward and his mother returned from being with my father at breakfast time; I could see through the front window before they entered the house, that they were disturbed and angry. On entering the front door, my mother went right straight down stairs without looking my way and Edward stepped right up to me straight and firm like a soldier and said "I am not giving you any money". I said "I did not ask you for any". After those words were said he left and there were no good-byes between us when I was leaving. I did not know where Edward and his mother were. I did not go to see my father either before I left Vernon; I did not want to disappoint him by having to tell him that his Katharine and their son Edward did not carry out his wish. I was at a loss for words – even now, it still bothers me: did my father believe, during the remaining days of his life that I was so ungrateful that I did not acknowledge or thank him for remembering his boxcar girl. Yes! It does still bother me, and it will till the end of time. I was wrong: I should have told him. He always said and believed that honesty is the best policy. I did not respect his honour by not telling him.

The 1990's – Eugenia Irene moved to Vancouver Island April 30th 1990 and was not at my father's 85th birthday. In September of that year my father had Edward drive him and his mother to Vancouver Island; he was determined to inspect my new way of life on the Island; and he was pleased. He loved Campbell River from what he saw; he said there were more boats docked there from any boats he had ever seen anywhere, and that he hoped to return one day to Campbell River: to the Island. My only and 1st born son William with his family had their home in Ranch Park, Coquitlam. Joyce and Curt with their daughters Marie and Rhonda were living in their beautiful home in Kelowna near the Okanagan Lake. Marshall James and Jo, with their son Yuel and daughter Voya were in their home by the sea in Victoria. Patrick and Connie with their three sons were in their large remodeled home on their ten acres in Vernon. Henry and Rita; a dear heart and gentle people lady; were living in Vancouver – the Marpole area. Henry's first born son Marshall was in California, Matthew in Burnaby - Vancouver, Mark in the Cariboo, Prince

George area, Melvin in Vancouver and Michael a lawyer in Vernon. Henry's daughters: Lauraine, Laura and Linda were all in the Vancouver area. Edward with his fortune hunter Susan; whom he met entertaining the abandoned souls at the Polson Acute Centre, with her off-key songs, were in Vernon and too soon left to live in a beautiful two story home on acreage in Cherry Creek – between Kamloops and Savona. Edwards first born son Lorne was in Bangkok, Thailand. Clinton was in the Cariboo where he grew up; — the Quesnel - Prince George area. Clinton often travelled to Vernon to be with his grandfather; to visit him in his grandfather's home, and in the hospital. A thoughtful tall dark and handsome young man is Clinton, a pleasure to see and to be with. Edward's daughter Suzanne was in the Burnaby - Vancouver area. Anne now retired and living in a new "The Fraserview" apartment of her own in Maple Ridge, was enjoying her every day swimming in the new, modern Maple Ridge swimming pool; and the calm of a small country town. Anne's first born son Dana was in Vancouver – the West End area, and Keith was in Pitt Meadows – Port Coquitlam. Father said that over the years he had saved some money for his daughter Anne; that during the years 1939 – 1943 on the farm in Alberta, Anne was always a great help to her mother with the babies, in caring for them and for working in the home. And for the help they both received from Anne during and after the great move they made to the Okanagan in 1943. Anne said she never received her father's money gift of his love and appreciation: That her mother told her to tell her father that she did receive it even though she did not.

My father told me several times over how very pleased he was and how grateful to have had his first born grandson, William visit him. He said that his grandson William took him in a wheelchair out into the open air along the streets of Vernon and out into the park, away from everyone at the care centre; that they walked and they talked for several hours, just the two of them before he had to be returned to the centre. My father said William showed a kindness he appreciated and will remember forever.

For June 20, 1992, my father asked Edward to arrange a family, local friends, neighbours and acquaintances – no out of town guests, to join him and his Katharine at the Vernon Seniors Centre for a buffet lunch, a visit and music which he funded for all. This day, June 20, 1992 was their 66th wedding anniversary and his reason to bring everyone together. Father somehow knew that this would be their last anniversary to celebrate: And it was. After the Seniors Centre many people came to our family home and father was determined to stay up for as long as everyone else did. This was a warm quiet beautiful night and father sat in his wheelchair on the patio at the entrance door to the family room downstairs with his sons and all the men that were there visiting; sharing conversation as they all sat around the picnic table and talked until midnight. Marshall James returned his father to the Polson Care Centre after midnight on this night of his 66th wedding anniversary.

The month of July 1992 sneaked up on us with the weeks passing by very quickly. My father's strength began to rapidly deteriorate. He did not seem to have the will to go on. He was so alert and so strong on the day of his 66th wedding anniversary, I was not aware that he so suddenly would stop his fight: In July 1943 he put up a real fight to move his family of nine to his chosen sunny Okanagan Valley. That was 49 years ago when he was 38 years old and today; this Tuesday, July 28, 1992 was a sunny Okanagan Valley regular work day:

45th Wedding Anniversary 1971

50th Anniversary 1976

ANNIVERSARY

Mr. and Mrs. William Yawney celebrated their golden wedding anniversary at a party attended by more than 150 guests at the Vernon Lodge recently. The couple, who came from Alberta to Vernon 35 years ago were greeted by all the members of their family when they arrived, including their seven children and 18 grandchildren. On their table was a wedding cake and a specially-framed photograph taken at their wedding 50 years ago. Mrs. Yawney was escorted into the banqueting room by her grandson, B.C. Lions football star Bill Norton, and her husband came in with his eldest granddaughter, Lorraine Yawney.

(Daily News photo)

A Vernon friend, sister Jenny, sister Agnes, Katharine, William, Bill Halina, Julianna's son John, George Creston

60th Wedding Anniversary - William, Katharine, children and grandchildren

65th Wedding Anniversary 1991

66th Wedding Anniversary 1992

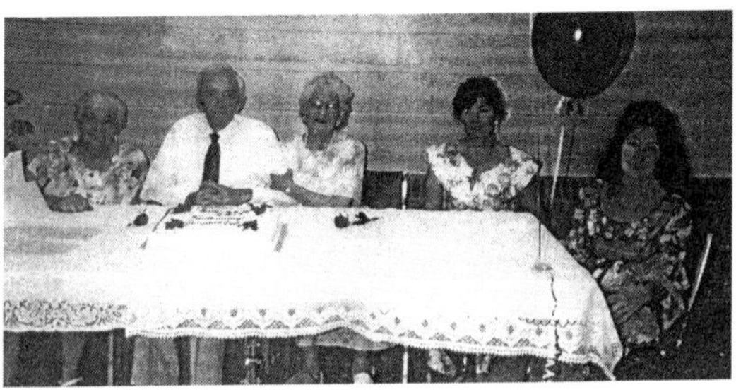

Patrick was on his way to Revelstoke to his job driving along the highway; as he arrived in Armstrong – father's first stop before he entered the Okanagan Valley in 1943 – something made Patrick stop and turn back; that he must go back to see his father. And he did. He was still with his father when Joyce arrived from Kelowna before noon. Patrick then walked away from his father's bedside and that was his very last visit with him. Joyce sat beside her father, holding him and together they were singing her father's favourite mountain song, the song his father used to sing while building the Canadian Railroad. Marshall James was at the family home talking on the telephone to doctor Hugh Clarke; asking the doctor if he may take his father to his family home for that afternoon. Edward with his mother were at the funeral home making arrangements. My father must have known that that was where they went and did not want to disappoint them by still being there when they return. Henry was in transit, and Anne in Maple Ridge was arranging her flight to Vernon. A whole month had passed since she last saw her father and this day she was returning to be with him again. July 28, 1992 my R.V. van was packed; I was ready to leave the Island: First I went into town to pick up some food to have with me while travelling. Returning just before noon: When I turned my van onto the road towards my home an unbelievable sight appeared before me in the sky; a jet trail formed a perfect cross. This jet trail cross was so perfect and so unbelievable I was stunned. I stopped my van in the middle of the road in amazement and disbelief and I am not sure for how long I stayed there staring at that perfect jet trail cross before me. The jet trail started narrow at the road I was on and gradually widened up high in the sky and stopped. Another jet trail crossed the upright jet trail to make a perfect cross. It did not fade away. Confused,

I went home to water my garden and flowers. It was 4:00 p.m. in the afternoon when I was leaving to start driving the highways when my telephone rang: It was my brother Marshall. He said my father passed away at noon. At noon while I sat in my van in the middle of the road staring at that jet trail cross in the sky.

Joyce was the only one beside her father, holding him, when he said to her "I am going to go now" and together they kept singing

У горах Карпатах
Ої, Там би я жив
Згори на долину
Все би м ся дивив – when in:

His pale face and in his eyes though now he could not see was a sadness and guilt that he was leaving. Leaving his Katharine, his love, his life, his partner of 66 years whom he cared for with all his heart, worked with and walked this world with;

Quietly tenderly he murmured his parting words
as he was drifting off into the universe.
"Come walk with me to eternity"
Alone and afraid he left this world
The Summer of '92

The Last Word

At the chapel of the Vernon Funeral Home; all of Vernon and all of the Okanagan Valley were there on Friday, July 31, 1992. Plus many, many more from out of province. The surrounding streets and avenues, the chapel yard and the standing room in the chapel were packed full, with people who cared and who came: Who respected my father through the years. Most comforting to me was to see Bill Balash, the first born son of John Balash, who helped my father through the beginning years, getting his B-Y Store in progress. (That was the summer of 1945.) My son William was not there on that day of July 31, 1992. He was in Alabama, U.S.A.

Edward stood at the rear of the chapel seated area playing solo violin, with a river of his tears flowing down his violin as he played father's favourite dance tunes: Some of the tunes he played were – Red River Valley, Home on the Range, You are my Sunshine, Sail Away, Down in the Valley, Maggie, I'll Take You Home Again Kathleen, Smiles, and Now is the Hour.

Marshall James was the Officiant who together with Edward arranged the Hill Billy Service.

Anne – the oldest daughter read a poem:

"Goodbye with Love"

"You left us Tuesday and quickly so
I guess you knew it was time to go.
Better than those who would have to stay
For one more hour or another day.

It was time to leave, while you still could be
Though physically bound, in spirit free.

You still could smile and still could sing,
Still could do so many things.

There are some battles that can't be won
The fight was hard but now it's done.
You fought so well and gave your best
And now, dear Dad it's time to rest."

Anne's first born son Dana, had these words to say:

"I will always remember my grandfather's imposing presence. Physically he was a mountain of a man with a booming voice, barrel chest and firm handshake, all of which he used with great effect to ensure that my cousins and I were kept firmly in line.

Intellectually, he could hold his own with anyone (he was very much a graduate of the "School of Hard Knocks"), he could always be counted on for a rousing and challenging political debate. And while he might not agree with your position or perspective, he allowed you the right to believe what you chose to believe.

His greatest contribution to my life though was his ardent belief in social responsibility. He felt that we should all work towards creating a society in which there was no hunger, no poverty, no destitution, a society where shelter, medical care and education were available to all. I found his passion all the more remarkable because in my mind he was the "self-made" man; he had built a successful business, raised a good family and become a prominent member of his community.

My grandfather encouraged us to be the best that we could be. The tremendous artistic, academic, athletic, professional and personal achievements of my aunts and uncles and cousins reflect this. I am incredibly grateful to have had this influence on my life."

Henry's son Michael – *held his young son as he spoke of his*

grandfather's 5 generations in Canada.

My son William's telegram read:

"In the same manner that great political leaders influence a nation, my grandfather William Yawney influenced his family through leadership. His philosophies like the great politicians aren't compatible with anyone's personal ideals. However, if we have taken one single point of this unique man's teachings, utilize the wisdom in a positive manner, he will live on forever in your hearts as he has and will in ours.

My grandfather will always be missed but not forgotten."

Marshall James – at graveside, had these words to say:

"The great sense of loss and the deep sorrow we all feel, words alone cannot express. Parting with you Dad, has left a painful void in our lives that will challenge our strength and resolve to carry on in your footsteps. We know that it would be your greatest wish that we who remain should try to understand better the transient nature of life and thereby truly cherish all the precious moments we have together. We must strive to honour your wish.

For us, there will always be a very special Dad. You know you will ever be loved by us. Our best memories of you will remain in our hearts for all time. May you rest long and in peace."

A reception followed, at the Vernon Seniors' Centre, with a buffet lunch and champagne catered to all the guests who were there to help us through this day. The Okanagan Valley Old Time Fiddlers provided the music. They all came; they all wanted to be there with us, and they played with energy from the heart, through to the end.

That evening after the sun went down, as we all sat on my father's sundeck; a symphony of fire was exploding behind the mountains just south of the Silver Star Mountain. The mountain my father used to take his family for sleigh rides in the snow through the woods at Christmas time. The mountain he walked many, many times with his sons when they went hunting. The light did not scatter as the symphony of fire: It was a solid red-orange light that flashed on and off for quite a long time. Then came the dark of night. We all stayed out on that sundeck looking up at the twinkling stars high up in the clear dark sky above with the Vernon city lights beyond and out of reach reminding us of the days gone by. The night seemed so quiet and at peace when suddenly birds began to chirp, the twinkling stars disappeared and the dark endless hollow sky became clear. Morning had broken and a new day was born.

All the dear hearts and gentle people of my father's Sunny Okanagan Apple Valley will remember; and forever he will remain.

Dear To My Heart

Walking With Dad

Oh, what rich dividends it yields
To walk with Dad in woods and fields;
To pause and rest in shady dells
Where mother nature's wonder dwells.
To simply wander here and there
And breathe the fragrant summer air;
To hear the brooklets gentle flow
Where Dad had fished so long ago;
To see a place where wild ducks nest,
And sunlight shine on mountain crest;
Then as the shadows gently fall
To hear a whippoorwill's soft call
And watch a twinkling star respond
To shine above the silvery pond.
Yes, every child is proud and glad
When he goes walking with his dad!

Reginald Holmes